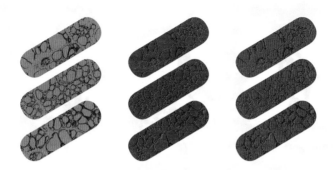

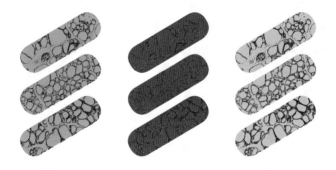

OXFORD BIOLOGY PRIMERS

Discover more in the series at
www.oxfordtextbooks.co.uk/obp

Published in partnership with the Royal Society of Biology

HUMAN INFECTIOUS DISEASE
AND PUBLIC HEALTH

≋ OXFORD BIOLOGY PRIMERS

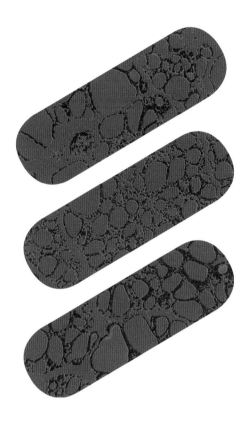

HUMAN INFECTIOUS DISEASE
AND PUBLIC HEALTH

Dr William Fullick MBBS MRCGP
Edited by Ann Fullick
Editorial board: Ian Harvey, Gill Hickman, Sue Howarth

OXFORD
UNIVERSITY PRESS

Royal Society of
Biology

Great Clarendon Street, Oxford, OX2 6DP,
United Kingdom

Oxford University Press is a department of the University of Oxford.
It furthers the University's objective of excellence in research, scholarship,
and education by publishing worldwide. Oxford is a registered trade mark of
Oxford University Press in the UK and in certain other countries

Published in the United States of America by Oxford University Press
198 Madison Avenue, New York, NY 10016, United States of America

British Library Cataloguing in Publication Data
Data available

Library of Congress Control Number: 2018959891

ISBN 978-0-19-881438-2

Printed in Great Britain
by Bell & Bain Ltd., Glasgow

PREFACE

Welcome to the Oxford Biology Primers

There has never been a more exciting time to be a biologist. Not only do we understand more about the biological world than ever before, but we're using that understanding in ever more creative and valuable ways.

Our understanding of the way our genes work is being used to explore new ways to treat disease; our understanding of ecosystems is being used to explore more effective ways to protect the diversity of life on Earth; our understanding of plant science is being used to explore more sustainable ways to feed a growing human population.

The repeated use of the word 'explore' here is no accident. The study of biology is, at heart, an exploration. We have written the Oxford Biology Primers to encourage you to explore biology for yourself—to find out more about what scientists at the cutting edge of the subject are researching, and the biological problems they're trying to solve.

Throughout the series, we use a range of features to help you see topics from different perspectives.

Scientific approach panels help you understand a little more about 'how we know what we know'—that is, the research that has been carried out to reveal our current understanding of the science described in the text, and the methods and approaches scientists have used when carrying out that research.

Case studies explore how a particular concept is relevant to our everyday life, or provide an intimate picture of one aspect of the science described.

The bigger picture panels help you think about some of the issues and challenges associated with the topic under discussion—for example, ethical considerations, or wider impacts on society.

More than anything, however, we hope this series will reveal to you, its readers, that biology is awe-inspiring, both in its variety and its intricacy, and will drive you forward to explore the subject further for yourself.

ABOUT THE AUTHOR

Dr William Fullick MBBS MRCGP trained at the University of East Anglia, qualifying in 2009. He went on to complete his junior doctor years in Devon and Cornwall, rotating through a number of different medical specialities before deciding on general practice as a career and passing his Royal College of General Practitioners exams in 2015. Working part-time as a GP in the West Country allows him to see a wide variety of patients, whilst also fitting in his lifelong passion for writing. William really enjoys using his medical and scientific knowledge to educate and inspire others, helping people understand more about their own bodies, the diseases that beset them, and the treatments available. When he's not seeing patients or writing, he enjoys walking the cliffs, kayaking, and stand-up paddle boarding.

ACKNOWLEDGEMENTS

In writing this book, I am indebted both to Milly, my fiancée and now wife, and to my mother. Your enduring support and encouragement has allowed me to pursue my passion for writing. Thank you.

CONTENTS

ABBREVIATIONS

cm	centimetre
kg	kilogram
nm	nanometre

1 THE SILENT ENEMY

From the earliest days of humanity, we have been shadowed by unseen enemies. Invisible to the naked eye, they have silently stalked us for thousands of years, causing death and disability to untold millions around the globe. We cannot smell them, taste them, or scare them away—and for the overwhelming majority of human history we have been utterly powerless to stop them. We have watched as they killed or maimed the young, the old, and sometimes even the very fittest and healthiest in a population, able to do little more than comfort the victims and hope that they might be strong enough to fight off their attackers. We have carried our killers with us everywhere we have travelled—in our bodies, on our clothes, in our food and water supplies, and with our animals. As our species has grown and developed, so too has this threat. Our enemies have become more efficient at replicating and dispersing themselves, even evolving defence mechanisms against the very weapons we have discovered to combat them.

Despite the incredible developments in modern medicine over the past century, these ancient enemies still kill millions of people every year. In 2014 the World Health Organization (WHO) stated that this threat has 'the potential to affect anyone, of any age, in any country'. It is almost certain that everyone who picks up a copy of this book has been a victim, although you may not have realised it at the time. Perhaps you had a stuffy nose, a dry cough, or some diarrhoea and vomiting. It's very probable that you carried on with your life much as usual—maybe you felt a little under the weather for a few days but recovered, as humans tend to do. It's possible you were more unwell than this—maybe you saw a doctor, took a course of antibiotics, or even needed a stay in hospital.

The invisible enemy is, of course, infectious disease. Caused by bacteria, viruses, fungi, parasites, and prions, infectious diseases kill around fifteen million people globally per year, although in countries like the UK and the USA we have become quite good at dismissing their potentially devastating effects. 'It's just a touch of flu', we say, perhaps forgetting that in 1918–19 'Spanish flu' killed over twenty million people. Even in more economically developed countries, infections diseases held a terrifying grip only a few generations ago—you can see an example of this in Figure 1.1. With modern medicine, many previously fatal infectious are either entirely avoidable (through vaccinations) or treatable (using medicines like antibiotics). It is now rare for a child to die of scarlet fever, measles, or whooping cough in the Western world. However, in poorer parts of the globe, millions of children still die from infectious diseases every year.

Figure 1.1 A mother's name surrounded by those of several of her children—she died during childbirth, and one of her children died whilst still very young. This simple headstone serves as a chilling reminder of an age before infections could be easily treated, when death in childhood was considered normal.

William Fullick

The evolution of infectious disease

Where have infectious diseases come from? To fully understand the origins of the pathogens that plague us, we must travel back in time to the emergence of life on Earth, approximately 3.8 billion years ago. Our current model of the origins of life suggests the first living things were very simple, probably based on single RNA strands rather than the DNA that makes up the genetic code of most organisms today. There is no evidence in the fossil record for these organisms, and their existence has been much debated among the scientific community, but it seems likely that these 'first replicators' existed, and eventually led to the evolution of more complex, DNA-based organisms. The first evidence in the fossil record for prokaryotic organisms—the early ancestors of modern bacteria—occurs about 3.5 billion years ago. At that time, we think the atmosphere of the Earth was made up of a mixture of nitrogen, hydrogen, carbon dioxide, and water vapour, with little oxygen—very different to how it is today. Nevertheless, there are fossils from this era which seem to provide evidence of single-celled life. Fast-forwarding to modern times, there are now more types of microorganism than there are other living things. Most of them have absolutely no impact on the human race. Some are beneficial—we use them to degrade waste products, to help us in making food or drink, within our bodies to help digest our food and balance our metabolism, and even to produce some of our antibiotics. A tiny number of them cause disease.

Billions of years of evolution led not only to many different species of microorganism, but also to multicellular life, from simple sea-dwelling sponges, to vertebrates, and eventually the gigantic reptilian creatures known as the dinosaurs. About sixty-five million years ago, the dinosaurs were wiped out in the Cretaceous Extinction, heralding the rise of the mammals. Many millions of years later *Homo sapiens*—modern humans—finally took their first tentative steps. From an evolutionary perspective, we have only existed for a very short time indeed—a mere 200,000 years or so—and during that time we have advanced incredibly quickly. From small family, and then tribal, groups, humans began living in closer and closer proximity to one another, first in villages, then towns, and eventually cities. We developed agriculture, writing, religion, and legal systems—all the trappings of a so-called 'civilised' society. We domesticated animals—both for food and for companionship. We were no longer just occupied with finding our next meal and reproducing. As the amount of food we produced and stored increased, we freed up enough time and spare resources for the creation of art, of entertainment, of social welfare. We began to care for the old and the sick, who otherwise would have died. But these steps forward did not come without a price, and infectious diseases caused by bacteria, viruses, and other pathogens accompanied humanity as we advanced. Many of the changes that allowed ancient cultures to flourish also permitted infectious disease to spread more easily than ever before, and exposed humanity to diseases which they might not have encountered in a more scattered existence. Case study 1.1 gives you a recently discovered example of this process in action.

In addition, the cats and dogs we love as pets and the livestock we farm and eat all carry their own pathogens, from bacterial and viral infections to infestations of worms and parasites. As people domesticated these animals, living and working closely with them, it became easier for diseases to cross the species barrier. Animal parasites—fleas, ticks, and mites—as well as biting insects such as mosquitoes—all have the potential to carry and spread disease. Over time, these evolved to affect humankind. Cramped living conditions in towns and cities allowed infections to spread more easily from person to person through close contact, droplet spread, and vectors such as fleas and lice. Contaminated water sources and a poor understanding of basic hygiene led to outbreaks of diarrhoeal illness such as typhoid and cholera. Sexually transmitted infections (STIs) could threaten larger populations, as greater numbers of sexual partners became available. Individuals became reservoirs of infection, passing diseases to those around them, who would in turn infect others. Epidemics spread through entire towns. Eventually, with improvements in travel allowing people to visit other cities, and then other countries, it became possible for an infectious disease from a single source to affect thousands of people. The 'global village' is also a global melting pot of infectious diseases. Written records from the ancient world show that epidemics (known by the people at the time as plagues) ravaged the great civilisations of China, Greece, and Rome on many occasions.

Case study 1.1
Silken parasites: proving the spread of disease in the ancient world

The Silk Road is the term given to the ancient network of land- and sea-based trade routes between the Eastern world and the West. Stretching from Japan through to the Mediterranean Sea, the Silk Road permitted a cultural exchange between East and West, allowing for the transportation of many exotic goods—including the silk for which the route was named. You can follow the route in Figure A.

Figure A The ancient Silk Road trade routes, which allowed people and diseases to spread across the ancient world.

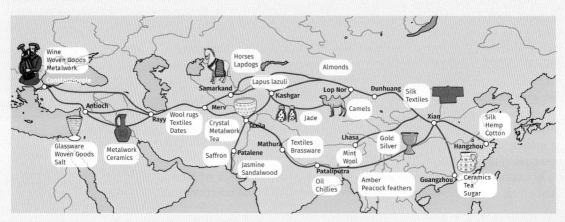

As different civilisations came together, ideas were also traded, with religions such as Buddhism, Jainism, and Christianity spreading throughout Eurasia thanks to the Silk Road. Buddhist monasteries set up along the roads provided welcome havens for weary travellers, as the route was long and dangerous. These monasteries would have given people a chance to rest, to eat, and to relieve themselves in comparative safety—but scientists have recently discovered that these ancient merchants may have brought more than silk along with them . . .

Excavations of an ancient **pit latrine** located at a relay station in the Tamrin basin revealed several 'personal hygiene sticks', which would have been used by travellers along the Silk Road to scrape away faeces from the anus. The sticks were then discarded into the latrine, where they lay undiscovered for 2,000 years. By examining fragments of dried faeces stuck to these sticks, a research team from Cambridge University discovered the eggs from four different species of parasitic worms—**roundworm**, **whipworm**, **tapeworm**, and **Chinese liver fluke**—you can see these in Figure B. This proved that people travelling the ancient route were infected with various different parasites—but the types of parasite were what interested the research team the most.

B Whipworm (*Trichuris trichiura*), roundworm (*Ascaris lumbricoides*), Chinese liver fluke (*Clonorchis sinensis*), and tapeworm (*Taenia sp.*)—the four different species of parasite discovered in ancient faeces on the Silk Road.

Helminths

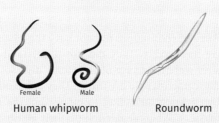

Female   Male

Human whipworm

Roundworm

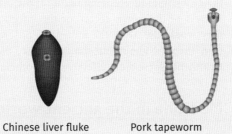

Chinese liver fluke

Pork tapeworm

Timonina/Shutterstock.com

The Chinese liver fluke is a parasitic flatworm, which causes abdominal cramps, diarrhoea, and jaundice in infected individuals. Eventually it can lead to liver cancers resulting from chronic, long-term inflammation. But the flatworm needs damp, marshy areas to complete its lifecycle, and the Tamrin Basin is an arid desert. The liver fluke must have come from somewhere else. Examining where Chinese liver fluke is found, the researchers realised that this infection had almost certainly been carried over 1,000 km—and possibly as far as 2,000 km—by its human host.

This realisation was revolutionary. Other scientists had previously suggested that diseases such as **bubonic plague**, **anthrax**, and even **leprosy** had been transmitted along the Silk Road, but finding proof was difficult. Most bacterial infections leave little trace once they have infected an individual, but parasites will often deposit very durable eggs—which can lie dormant for many years. The relay station was used from approximately 109 BCE to 111 CE, so these eggs had been undiscovered for 2,000 years, but were still recognisable. Here was hard evidence that infectious diseases had been carried huge distances by travellers along the ancient trade routes, to places many thousands of miles from the original source of infection.

❓ Pause for thought

Think about other places where civilisations crossed paths throughout history. Are there any other obvious examples of the spread of infectious diseases that you can think of? Where is the scientific evidence for this? Can you think of any methods that could confirm whether one strain of pathogen is related to another?

Ancient civilisations did not have the advantages of the medical knowledge we have today. They did not understand that microorganisms and parasites were the cause of infectious disease, and they had no way of seeing these microscopic life forms. They could observe people becoming sick, and the sickness spreading to others, but the cause of the disease remained a mystery. When people fell ill, their ability to treat disease successfully was limited. Often the only option was to keep the patient comfortable whilst their own immune system fought off the infection. The patient would either recover of their own accord or the infection would overwhelm them, and they would die. Over the millennia, different civilisations developed their own theories about infectious disease, along with different methods of treatment—some more effective than others.

Medicine in the ancient world

Many of the early explanations of disease were based around spiritual or religious ideas and fears, with a priest, holy man, or shaman acting as a healer. In the earliest civilisations, or societies with no written records, it can be very difficult to track how diseases were viewed or treated. We do, however, have evidence from the past few thousand years of the theories about, and treatments for, infectious disease from many different cultures. Often quite different in their rationale, they frequently share similar traits and theories—as you will see throughout the rest of this chapter.

Ancient India

The Vedic period, or Indian Iron Age, lasted from about 1500 to 600 BCE. There was a big migration of Indo-Aryans into India at the time, and we know about their ideas from a number of ancient texts. These show that Vedic medical principles were based around a combination of spiritual beliefs and a complex knowledge of local plants, animals, and minerals. Schools of herbalists developed, and it was stated that 'for a knowledgeable healer, the herbs rally together akin to an army of kings'.

The approach of the healers was not scientific, however, and many of their treatments involved magico-religious aspects. They believed, for example, that fever could be 'transferred' from a patient to a frog tied under the bed, and most healing preparations were accompanied by prayers or chants. Vedic principles did identify some important aspects of infectious disease, stating that fever was a 'sister' of certain other conditions. They also considered infestation with worms and illnesses caused by 'unwholesome or contaminated foods' as contagious.

The Vedic system was replaced by the principles of Ayurvedic medicine from about 600 BCE. Much of the evidence of Ayurvedic medicine comes from observations from Greek visitors to India at the time, and from Sanskrit texts dating from the common era (CE). Ayurvedic medicine attempted to shrug off the magico-religious aspects of Vedic medicine, and instead turned towards logical reasoning and knowledge of the medicinal effects of plants and other substances. It describes a systematic approach to examining and diagnosing a patient—but still describes the causes of diseases as being due to disorders of the humours, a common theme in older approaches to medicine (see 'Medicine in the Greco-Roman world' later in this chapter). It is certainly possible—although far from certain—that some of the herbs used by the Vedic and Ayurvedic doctors could have had antimicrobial properties, and so some of their treatments may have been at least partially effective at treating infectious diseases.

Traditional Chinese medicine

Doctors in ancient China developed a complex understanding of health and disease, including infectious diseases. At some point between 25 BCE and 220 CE a doctor named Zhang Jong Jing wrote a book entitled *Shang Han Lun* ('*Treatise on Febrile Disease*'), describing the symptoms of various 'stages' of infectious disease, and prescribing treatment options based on whichever stage the patient displayed. He noted that feverish conditions could spread from person to person 'via heaven or earth' (via airborne routes, or by contact from infected patients) and could therefore be transmitted 'from one person to the entire household, from one household to the entire street, and from one street to the entire village'. Traditional Chinese medicine (TCM) used a vast array of herbs and herbal preparations to treat illnesses according to these principles. Many of these are now recognised to have antimicrobial properties, so Chinese doctors may well have experienced a degree of success in treating patients with certain infections. Many of the herbal preparations described in the ancient texts of TCM are still used in Chinese medicine today—have a look at Figure 1.2 to see some of them.

Figure 1.2 Millions of people still use traditional Chinese medicine today, in China itself and around the world.

iStock.com/marilyna

Public health also became a concern as the population of China grew, with diseases such as smallpox spreading quickly in crowded cities and affecting thousands of individuals. Government officials during the Song Dynasty (960–1279 CE) started to enforce basic public health measures in an attempt to slow the spread of infectious disease. Road sweepers and 'night soil men' (collectors of human waste) were employed to keep the streets free of rubbish and faeces, and spittoons were provided for people to spit into. Public dispensaries, hospitals, and orphanages were set up in urban areas. This demonstrates that doctors and officials clearly understood the transmissible nature of infectious diseases, even if the specific cause of them had not yet been identified. Although smallpox was untreatable once contracted, these basic public health measures may have helped to reduce its spread—and perhaps also the spread of some other contagious illnesses.

Medicine in the Greco-Roman world

Both the Greeks and the Romans were well aware of the effects of infectious diseases, with both civilisations suffering from plagues at various times during their existence. They both also attributed disease to 'disorders of the humours'—an imbalance of black bile, green bile, phlegm, and blood—as well as a belief that many illnesses could be attributed to having attracted the disfavour of the gods. It was also theorised that social class, gender, and geographical location could predispose an individual towards certain diseases. Some Greek and Roman physicians (such as Hippocrates and Galen) scorned the idea of supernatural events causing disease, and instead suggested that infectious diseases were contracted due to the poor quality of food, air, or water to be found around the patient. Contemporary Hippocratic texts describe a wide range of infectious diseases which are clearly recognisable today, including meningitis, upper respiratory tract infections, pneumonia, and even skin infections. However, the true cause of these infections remained hidden from the physicians of the Roman and Greek worlds, because glass-working was not advanced enough to produce magnifying lenses of sufficient power to view microorganisms.

Medieval Europe

In Europe, ideas about infectious disease initially followed similarly spiritual lines to India and China.

Medieval doctors tended to use the works of the Roman and Greek physicians as their basis of understanding for disease, so the ideas of 'humoral imbalance' and diseases being caused by 'foul air or water' (known as miasma theory) persisted in Europe throughout the Middle Ages. From texts at the time, it is clear that people understood that disease could be passed from person to person, but that the exact mechanism of this was very unclear. This period of history saw further epidemic diseases ravage the peoples of Europe– perhaps most notoriously the bubonic plague, known as the Black Death for the black swellings which formed on the bodies of its victims. Outbreaks of this terrible disease are estimated to have killed up to 60% of the population of Western Europe. We now know that bubonic plague is caused by the bacterium *Yersinia pestis*, carried by rats and transmitted by the fleas which live on them, but at the time the cause was thought to be miasma. People carried bundles of herbs and spices with them, which they believed would protect them from the foul airs which caused the plague. As you can see in Figure 1.3, doctors embraced this idea with enthusiasm!

Those who were infected were quarantined, showing that there was some awareness that the disease was contagious (as Case study 1.2 clearly demonstrates), but without proper public health measures and a sound understanding of the disease process itself, the plague spread through Europe, with very little to hold it in check. Without any effective treatment, bubonic plague was rapidly fatal, killing over half of those it infected. Measures were often put in place by local authorities to avoid handling the bodies of the dead as far as possible. The belongings of victims were also often destroyed in an attempt to avoid contamination. Many of the dead were buried in large, unmarked 'plague pits', which are sometimes still discovered today when building works take place in European cities.

Occasionally scientists or scholars would suggest alternative theories for infectious disease, and these were sometimes surprisingly accurate. Fracastorio, an Italian poet, physician, and scholar, suggested in 1546 that infectious diseases were caused by invisible infectious spores, which he called *seminaria* ('seeds'), although he stated that these were caused by the alignment of the planets. This idea gained some traction in its time, but it would be another hundred years before microorganisms were first seen using a microscope, and a further two hundred years before it became widely accepted that these 'seminaria' were the cause of infectious disease.

Figure 1.3 Doctors during the Black Death may not have known what caused the plague, but they knew that it was contagious. This terrifying costume was designed to shield its wearer from the plague-causing miasmas. The beak could be stuffed with pleasant-smelling herbs and spices.

Wellcome Collection

Case study 1.2
The plague village of Eyam

Eyam is a pretty village in Derbyshire, UK with a dark story buried in the village churchyard (Figure A). The picturesque streets and ancient church conceal a story of death and selflessness dating back to the 1600s. In 1665, a parcel of cloth arrived for the village tailor. This cloth had been sent from London—which was gripped by the epidemic of bubonic plague that was sweeping through Europe at

Figure A The bubonic plague inflicted a grim toll on the inhabitants of seventeenth-century Eyam.

Steven Gillis/Age Fotostock

the time. What the tailor did not know was that the cloth was infested with fleas, carrying with them the plague bacterium *Y. pestis*, shown in Figure B. Within a week, the tailor's assistant—a man named George Vicars—had succumbed to the disease. Others within the tailor's household began to suffer and die shortly afterwards, and the outbreak began to spread throughout the village.

Figure B *Yersinia pestis*, the microorganism responsible for causing the bubonic plague.

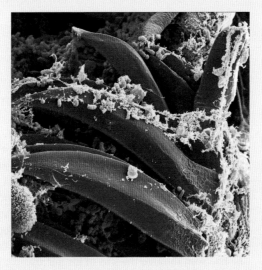

Public Health Image Library (PHIL)/CDC/National Institute of Allergy and Infectious Diseases (NIAID)

In desperation, the villagers turned to their religious leaders—William Mompesson and Thomas Stanley—who introduced a number of measures to try to contain the spread of disease within the village. They insisted that only family members could bury the bodies of those killed by the plague—reducing the number of people coming into contact with infected bodies. They moved church sermons to a nearby outdoor amphitheatre, allowing people to keep a safe distance from one another. Most dramatically, they also decided to quarantine the village itself, isolating themselves from the surrounding towns and villages. A series of 'plague stones' were designated at the boundaries of the settlement. The Eyam villagers would place money soaked in vinegar—believed to kill the infection—onto these stones, in exchange for food or medicine from the other villages in the area. The stones also marked the village boundaries, and were forbidden to be crossed, either by villagers or outsiders.

The quarantine was certainly effective. The other measures to control the plague within the village were perhaps less so. Within eighteen months, only eighty-three people remained alive out of an original population of about 350. The plague had killed seemingly at random—some of the survivors had buried many of their own family members but had never displayed symptoms themselves. With the benefit of a modern understanding of infection, we can now assume that they perhaps had developed an immunity to the plague bacterium following exposure—or possibly that their own immune systems were simply strong enough to fight off the infection without them succumbing to it.

Eyam is now famous for the selflessness of the villagers in quarantining themselves away from the rest of the world. Many of the graves of the plague victims can still be seen in the churchyard and fields around the village.

 ## Pause for thought

With the benefit of modern knowledge of infectious disease, can you think of any other ways that the quarantined villagers of Eyam could have helped to reduce the spread of the Y. pestis *bacterium within their community?*

The discovery of microorganisms

Bacteria were first properly described in 1676. A Dutch draper named Antonie van Leeuwenhoek produced his own glass lenses and manufactured a crude light microscope, which could magnify to between 30 and 256 times larger than life size. This allowed him to see bacteria, along with the microscopic structure of many other things in the world around him. He examined and described the microorganisms present in rainwater, sea water, and well water, describing his findings in a letter to the Royal Society in London. This was followed by a second letter in 1684, in which he described various other things he had seen through the lens of his microscope, including the bacteria he had observed in the tartar scraped from his own teeth. These two letters contain the first examples of drawings of bacteria, and the letters themselves have survived through to modern times, giving scientists today a good idea about how bacteria were visualised hundreds of years ago. Have a look at Figure 1.4 to see some of Van Leeuwenhoek's original drawings. Robert Hooke, an English scientist, published observations of the first microscopic fungi at a similar time, although it would be a further century before many of the common microscopic fungi and moulds would be described and identified (by Micheli, an Italian botanist). These discoveries made surprisingly little difference to the way that the medical establishment thought about infectious diseases. The link between Van Leeuwenhoek's 'animalcules' and the many infectious diseases affecting humanity was not made for many years.

Figure 1.4 Van Leeuwenhoek's drawings clearly demonstrate some of the important characteristics of various different types of bacteria we recognise today, including bacilli and streptococci.

Wellcome Collection

Debunking the myth of spontaneous generation

Although the existence of microorganisms was demonstrated by Hooke, Van Leeuwenhoek, and others during the seventeenth century, the question of where they had originally come from had yet to be answered. Most people at the time believed that all life could originate from other matter, either dead or living. It was thought that rotting flesh spontaneously produced maggots, that fleas arose from dust, and that even complex organisms could be created with the right 'recipe'. Van Helmont, a seventeenth-century alchemist, published a recipe he claimed would create mice! The first serious attack on this theory came in 1668, nearly a decade before Van Leeuwenhoek described bacteria. An Italian doctor and poet named Francesco Redi believed that maggots were not spontaneously generated by rotting meat, but rather were the products of eggs laid by flies. He proved this by conducting an experiment in which he laid out various samples of meat, some covered, some in sealed flasks, and some open to the air. As he had predicted, maggots only occurred in the samples open to the air, as flies could not reach the sealed samples. Despite this rather compelling evidence (and an easily reproducible experiment), belief in spontaneous generation continued without any significant changes—even Redi himself still believed that spontaneous generation would occur under certain circumstances. In some ways, the invention of the light microscope and visualisation only served to increase belief in spontaneous generation. To view these new 'animalcules' all one had to do was place some hay in water for a few days before looking at what appeared to be new life, created seemingly from nothing.

The argument over spontaneous generation dragged on for decades, and eventually, in the mid-nineteenth century, the French Academy of Sciences offered a prize for anyone who could settle the debate once and for all. A young French scientist named Louis Pasteur conclusively proved that microorganisms did not occur through spontaneous generation, using a technique which you can read about in Scientific approach 1.1. His published works were awarded the prize set by the Academy, but it would take a further twenty-five years for the wider scientific establishment to change their views, and proponents of the spontaneous generation theory survived even into the twentieth century.

Scientific approach 1.1
Pasteur and the swan-necked flasks

In order to disprove the theory of spontaneous generation, Louis Pasteur (see Figure A) needed to devise an experiment which clearly showed that microorganisms could not appear in a sealed environment. It had already been proven (by an Italian priest named Spallanzi) that boiling a liquid would kill any microorganisms within it, and Pasteur used this knowledge when devising his experiment. Have a look at the main stages of his method in Figure B.

Pasteur used these results to prove that, in a sealed environment where microorganisms had been previously eradicated, no further life could arise. This is a good example of using a scientific approach to test a hypothesis—in this case, that life could spontaneously arise in a sterile broth. If the theory of spontaneous generation was correct, even the broth in the sealed flasks would eventually have produced life. The sealed flasks remained clear of growth, however, and Pasteur could therefore confidently claim that spontaneous generation was a flawed theory.

❓ Pause for thought

With modern knowledge of microorganisms, can you think of any important considerations when conducting an experiment like this? How could you improve on Pasteur's methods to reduce the likelihood of erroneous results?

Figure A Louis Pasteur (27 December, 1822—28 September, 1895).

Wellcome Collection

Figure B Pasteur's classic experiment with the swan-necked flasks.

1. Pasteur took a series of flasks with long glass necks. Into each of these he placed a broth, known to culture microorganisms.
2. The broth was boiled to kill off any existing microorganisms, and the necks of the flasks were heated and bent into an S shape. Some were sealed and some were left open to the air, but the S-shaped bend in the neck prevented any microorganisms from entering the broth.
3. No organisms grew in the flasks—the broth remained clear.
4. Taking some of the flasks, Pasteur then broke the necks at a higher point, and tipped the broth until it came into contact with the curve of the S shape—where microorganisms had been able to collect. In these cases, the broth became cloudy, proving that it had been contaminated by microorganisms.
5. Pasteur then broke the necks off some of the remaining flasks, allowing the broth to come into contact with the outside air.
6. Given time, microorganisms grew in these flasks, and the broth became cloudy.

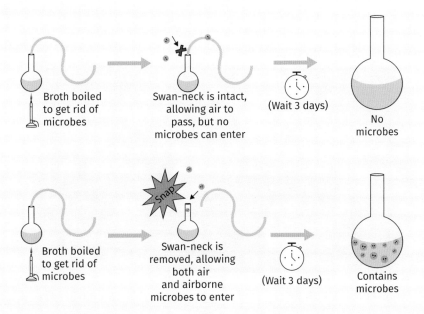

With permission of Cynthia Anderson Sanchez: http://www.scientistcindy.com

The germ theory of disease

The role of microorganisms in causing infections (known as the germ theory of disease) was hotly contested throughout much of the nineteenth century.

Case study 1.3
Ignaz Semmelweis, 'The saviour of mothers'

In 1846—some years before Pasteur's experiments—a Hungarian doctor called Ignaz Semmelweis started work in the obstetrics department at the Vienna General Hospital. You can see his portrait in Figure A. Semmelweis noticed that the death rates of women giving birth at the hospital varied tremendously between two different clinics. The main cause of these deaths was a condition called **childbed fever**—now known as puerperal fever. The first clinic (run by doctors and medical students) had an average mortality rate of 10%, whereas the second clinic (run by midwives) had a much lower average mortality rate—only 2–3%. This discrepancy troubled Semmelweis, and he set out to examine the differences between the two clinics.

After ruling out the climate, environment, and overcrowding—the climate and environments of the two clinics were identical, and the midwife-run clinic was always more crowded than the clinic run by the doctors—Semmelweis began to wonder whether the doctors were doing anything differently from the midwives. The breakthrough came about after the death of one of Sem-

Figure A Ignaz Semmelweis (1 July, 1818–13 August, 1865) washing his hands—a simple move that saved many lives.

Bettmann/Getty Images

melweis's close friends, who contracted symptoms similar to childbed fever after sustaining a puncture from a scalpel during an **autopsy**. Semmelweis immediately suggested a link between the **dissection** of **cadavers** and the high rates of childbed fever in the doctor-led clinic. The midwives who ran the second clinic did not perform autopsies and so avoided contamination.

Semmelweis suggested that the doctors washed their hands with chlorinated lime (calcium hypochlorite) when moving from the autopsy room to the obstetric clinic. This swiftly reduced the mortality rates of the doctors' clinic to equal that of the midwives' clinic, confirming Semmelweis's theory that 'cadaveric particles' were being transferred from the bodies in the autopsy room to the patients in the clinic, causing childbed fever and ultimately killing 10% of the women. He did not, however, make the link between these particles and microorganisms. He then moved on to two other hospitals, taking his methods with him. Using what he had learned in Vienna, he was able to dramatically reduce the mortality rates from childbed fever in these hospitals as well.

Although his work was accepted in the hospitals where he worked, Semmelweis's publications were rejected by the vast majority of doctors and scientists at the time. Many doctors took offence at the suggestion that they should wash their hands between examining patients, claiming that they were gentlemen and therefore their hands were clean. There was even a religious argument against this life-saving change, with some arguing that women were supposed to suffer in childbirth, as Eve had suffered in the Bible.

Semmelweis became more and more obsessed with his theories, and increasingly angry towards those who would not accept them. He wrote a series of aggressive and hate-filled letters to many of the leading obstetricians of the time, denouncing them as 'murderers' and 'ignoramuses'. His mental health declined, and he was admitted to an asylum for the insane in Vienna. Two weeks later, in an ironic twist of fate, Semmelweis was dead at the age of forty-seven. His cause of death was **sepsis**, thought to have occurred after a beating from the guards at the asylum.

It would take several decades before Semmelweis's contribution to infection control and public health would be recognised—during which time the mortality rates in obstetric clinics remained high. His findings are still important today, and good hand hygiene remains an important part of modern medicine. When this is neglected, it can lead to outbreaks of infections such as **methicillin resistant staphylococcus aureus (MRSA)**, and **norovirus** in hospitals and nursing homes.

❓ Pause for thought

Despite the fact that Semmelweis's theories were correct—and that he could prove that he had reduced mortality rates significantly—the scientific establishment refused to believe him, because his evidence was contrary to what many people at the time believed. We now call this 'the Semmelweis effect'. Can you think of any other examples of this throughout the history of science? Why do you think human beings will often reject new information, regardless of any evidence that supports it?

Germ theory moves on

Semmelweis had proved a link between contamination and disease when he showed that basic handwashing reduced rates of childbed fever (see Case study 1.3), but he did not make the leap between this and the germ theory of disease. Only after Pasteur's experiments in the late nineteenth century did doctors begin to suspect that microorganisms might be the cause of the infectious diseases which had been recognised for centuries. An English surgeon named Joseph Lister reasoned that he might be able to reduce rates of post-operative wound infections by using dressings soaked in chemicals to kill off airborne microorganisms. He chose to try carbolic acid, using the sweet-smelling substance not only to cover wounds, but also to sterilise surgical instruments, to wash his own hands in, and even to mist the operating theatre with prior to and during surgery. The results were remarkable, and post-operative wound infection rates dropped dramatically. The age of antiseptic surgery was ushered in, and the link between microorganisms and infectious disease was closer than ever to being fully understood.

Koch's postulates of disease

In the 1880s a German doctor by the name of Robert Koch proved that he could infect mice with **anthrax** bacteria from cattle, and that he could then isolate the same anthrax bacteria from the mice. He also proved that he could spread the infection from one mouse to another by inoculating a healthy mouse with blood from an infected one. Koch developed methods of culturing microorganisms that are still in use today, and published a document describing what came to be known as Koch's postulates—a set of instructions for determining whether or not a particular microorganism could be said to cause a certain disease. Simplified, these are:

1. The organism must be present in every case of the disease.
2. The organism must be isolated from a host of the disease and grown in pure culture medium.
3. Samples of the organism taken from the culture medium must then cause the same disease when inoculated into a healthy, susceptible animal in the laboratory.
4. The organism must be isolated from the inoculated animal and must be identified as being the same as the original organism isolated from the host and grown in pure culture medium.

Modern theories of infectious disease

Today, our understanding of infectious disease is more sophisticated than ever before—but new discoveries are still being made all the time. Viruses (many of which have only been identified in the past sixty years) are now recognised as important causes of many diseases such as influenza, chickenpox, herpes, and AIDS. Parasitic organisms also cause infectious diseases, particularly in the developing world. Many of these parasites—such as flatworms, tapeworms, and other helminths—are multicellular organisms. Others—such as amoebae—are single-celled eukaryotic organisms, differing from bacteria in crucial ways which affect how the diseases they cause can be treated. All are responsible for a significant amount of sickness and disability, with many parasites proving challenging to treat and even more difficult to eradicate entirely. Even more recently, scientists have identified microscopic folded proteins called prions. Whilst not 'living' in the same way that bacteria or parasites are, prions are able to transmit themselves and replicate their shape. In doing so, they fundamentally alter the structure of organic matter, causing a variety of illnesses in humans. We will examine these different pathogens further in Chapters 2 and 3.

As our understanding of the agents of disease has improved, so too has our knowledge about the ways our own bodies combat infection. We now know that many of the symptoms of infectious disease—fevers, increased mucus production, swollen tonsils, sore glands, or rashes—are often due to our own immune system combating an infection within the body. Fever, for example, has been shown to increase the mobility and effectiveness of leucocytes (white blood cells—a key part of our immune defences), as well as decreasing the effects of endotoxins produced by some bacteria. Some studies have shown that being able to produce a fever increases survival rates and decreases recovery time following an infection. We will look at the way that our bodies combat infectious disease further in Chapter 4.

Other symptoms of infection are now understood to be down to the direct actions of pathogens themselves. Many bacteria produce substances which are harmful to the human body, called toxins. Some of these—called endotoxins—form part of the cell membrane of Gram-negative bacteria. They are released only when the bacterium undergoes lysis (cell death). In diseases such as meningococcal septicaemia, a huge flood of these endotoxins can cause a sudden worsening in the patient's condition. The immune system responds to the toxin flood by increasing its inflammatory response, triggering the release of chemicals designed to aid the cells of the immune system in fighting infection.

Other compounds produced by some bacteria are called exotoxins. They can be released at any time in the bacterial life cycle. These exotoxins are broken down much more readily than endotoxins but are usually much more potent, with each exotoxin having a specific effect on the cells of the body. Our immune system can form antibodies against exotoxins, but the effects of the toxins themselves are often fatal before an efficient response can be organised by the immune system. A good example is botulinum toxin, released by the bacterium *Clostridium botulinum*, which rapidly acts as a paralysing agent, leading to death even after a relatively low dose.

Even if our immune system responds successfully to an infection, there can be long-lasting effects. We are only just starting to understand the links that exist between some common infections and other diseases such as cancer. A good example of this is a virus named human papilloma virus (HPV). There are over 100 different subtypes of HPV. Some are associated with genital warts, which are sexually transmitted. However, some strains of HPV (called 'higher-risk' strains) are also associated with an increased risk in cancers of the cervix, penis, and anus. This is because the HPV virus infects the cells of these areas and fundamentally alters their genetic code, changing the way that the cells divide. This increases the likelihood of cancerous cells developing. It is important to note that not everyone who is infected with HPV will develop one of these cancers, and some people develop cervical, anal, or penile cancers without HPV infection—but infection with the HPV virus definitely increases the risk. Some throat cancers also seem to be due to HPV infection—possibly due to changing sexual practices over the past fifty years or so, although research into this topic is still ongoing. Other types of infection have also been associated with specific diseases—such as *Streptococcus bovis* and bowel cancer. It is clear that we have much more to learn about the complicated interplay between microorganisms and our own bodies.

Chapter summary

- Infectious diseases have evolved alongside humanity, taking advantage of changes in human society to help their spread from person to person.
- Many early theories of disease involved religious/magical constructs.
- In many early models, diseases were described in terms of an imbalance of the four humours.
- Some historical attempts to treat disease may have had a degree of success, as some of the herbs used in traditional medicines have antimicrobial properties, and people began to understand the principles of quarantine.
- An understanding of the nature of infectious disease developed over time—linking the discovery of microorganisms to the theory of infectious disease was a slow process, taking many different scientists many hundreds of years to develop.
- The modern understanding of infectious disease has become increasingly complex, with some infections being implicated in other diseases such as cancer, which may occur many years after initial infection.
- We are learning more and more about the complex relationship between pathogens and the human body, and despite advances in modern medicine, infectious diseases still affect virtually everybody alive on the planet today.

Further reading

1. Milton Wainright and Joshua Lederberg, 'History of Microbiology' in *Encyclopedia of Microbiology* (Academic Press, 1992). https://profiles.nlm.nih.gov/ps/access/bbabon.pdf

A fascinating short history of microbiology as a science, from the very earliest days of microscopy to the current crisis in antibiotic resistance.

2. Kenrad E. Nelson and Carolyn F. Williams, 'Early History of Infectious Disease'. http://www.jblearning.com/samples/0763728799/28799_CH01_001_022.pdf

A look at how ancient peoples viewed diseases, along with early attempts to control and track outbreaks.

3. Anu Saini, 'Physicians of Ancient India'. *J Family Med Prim Care* (2016) Apr–Jun; 5(2): 254–8. https://www.ncbi.nlm.nih.gov/pmc/articles/PMC5084543/

A summary of the medical principles used during the Vedic and Ayurvedic eras in India.

4. Transcript of an address given by Louis Pasteur at the 'Sorbonne Scientific Soirée', 7 April, 1864. http://www.rc.usf.edu/~levineat/pasteur.pdf

This is an address that Pasteur gave after his famous experiment disproving spontaneous generation—imagine how controversial this must have been at the time!

Discussion questions

1.1 Think about the influence of infectious diseases on a specific aspect of human history. How might people at that time have responded differently with a more advanced knowledge of infectious disease? How might this have influenced the world we live in today?

1.2 How might early civilisations have chosen which herbs they used to treat diseases? Find evidence of at least one example where herbal treatments were developed based on their effectiveness. Suggest reasons why the approach may not always have been so scientific.

1.3 Discuss how close we are—or are not—fully understanding the complicated relationship between microorganisms and our bodies.

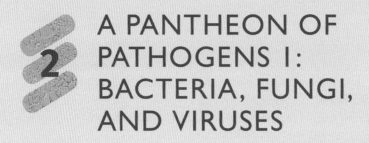

A PANTHEON OF PATHOGENS 1: BACTERIA, FUNGI, AND VIRUSES

This chapter will look at some of the many different **pathogens** which cause infectious disease. These pathogens include bacteria, viruses, fungi, parasites, and prions. In this chapter you are going to find out about the first three types of pathogens in this list—parasites and prions await you in Chapter 3!

Even microbiologists don't know how many species of bacteria, viruses, and fungi there are in the world, but the different types often share common features which can be used to identify them.

Discovering the structure of these pathogens (Figure 2.1) will help you visualise how they infiltrate the human body, understand how they can make us feel so ill, and show you the features used by our immune systems to identify and fight against them (which we will cover in Chapter 4).

Figure 2.1 Pathogens have many adaptations that enable them to invade and destroy our cells.

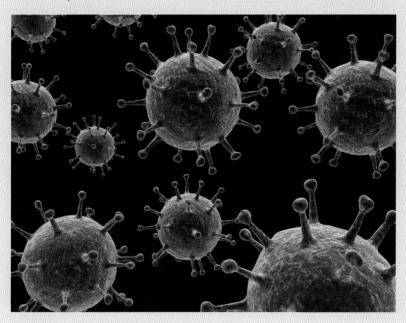

Sebastian Kaulitzki/Shutterstock.com

Bacteria—single-celled specialists

Bacteria are arguably one of the most successful forms of life on the planet. Look at almost any environment on Earth—from volcanic vents deep under the sea to the freezing Arctic tundra—and you will find bacteria, teeming in their millions. No matter how hostile a landscape may seem on the surface, under the microscope prokaryotic life will be thriving. They have even survived trips into space—frozen and seemingly lifeless amongst the stars, but quickly re-animating once they return to a more hospitable environment. The phenomenal success of the prokaryotes is down to several different factors. Firstly, they reproduce asexually and incredibly quickly, often filling an agar plate within a few hours or days. They don't need to waste resources looking for a mate, or time gestating their young. The offspring are immediately ready to start reproducing themselves, with no need to reach sexual maturity. This means beneficial genetic traits can be passed on very rapidly, allowing natural selection to occur far more quickly than in organisms with longer lifespans and slower reproduction. Secondly, many bacteria can pass their genetic traits horizontally as well as vertically, using parcels of DNA called plasmids. This allows beneficial mutations to spread through a population very rapidly, without the need for reproduction. They can even be passed between different strains of the same bacteria. Look at Figure 2.2, which shows how this plasmid transfer occurs.

Figure 2.2 Transferring plasmids from one bacterium to another allows prokaryotes to pass on beneficial genetic traits to those around them. This helps them adapt quickly to their surroundings.

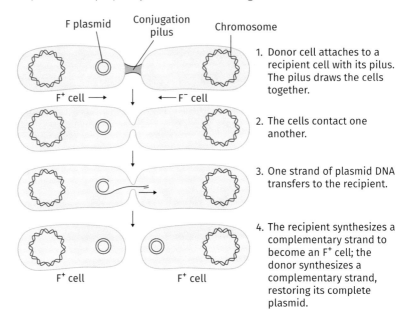

F plasmid Conjugation pilus Chromosome

F⁺ cell →   ← F⁻ cell

F⁺ cell   F⁺ cell

1. Donor cell attaches to a recipient cell with its pilus. The pilus draws the cells together.

2. The cells contact one another.

3. One strand of plasmid DNA transfers to the recipient.

4. The recipient synthesizes a complementary strand to become an F⁺ cell; the donor synthesizes a complementary strand, restoring its complete plasmid.

Thirdly, their simplicity and rapid dissemination of genetic material allows bacteria to become super-specialised, focusing solely on surviving in environments where other organisms would perish. This explains why bacteria can be found thriving in extremely low or high temperatures, or in places with oxygen levels so low that other lifeforms simply cannot exist. It is also why they have become so very good at infecting the human body. Of course, only a small number of bacteria have specialised in humanity. Most bacteria do not interact with people, either using other animals or plants as their hosts, or existing outside the bodies of other organisms. Many of those which do affect us are beneficial, such as the bacteria which thrive in our guts, helping us to digest food. These bacteria make up what is known as the human microbiome—up to 2kg of micro-organisms living in and on our bodies. Our understanding of the role of our microbiome is increasing all the time, as scientists discover that the bacteria are involved in processes from appetite control to the absence or presence of depression.

A very small number of bacteria have specialised in colonising human hosts. In order to do this, they must establish themselves in the area with the optimal environment, where they will either reproduce unchecked or eventually be destroyed by the cells of our immune system. The symptoms caused by this invasion either directly or by the immune response are what we call infectious disease. Read Case study 2.1—it describes how two scientists made the link between a type of bacterium and the diseases it causes.

Case study 2.1
Helicobacter pylori—a real stomach ache

In the earliest days of microscopy, many hundreds of different types of bacteria were described by scientists around the world. Because the links between bacteria and disease were not understood at the time, most of these discoveries were simply a description of the bacterium. This was certainly true in the case of *Helicobacter pylori*—an unassuming, spiral-shaped organism (Figure A), first found in the sediment of human stomach contents by a Polish scientist named Walery Jaworski. Discovering the bacterium in 1886, Jaworski did not attach any more importance to his find—and it would be nearly a century before the importance of *H. pylori* would be fully realised.

Inflammation of the stomach—otherwise known as gastritis—is a problem which affects hundreds of thousands of people around the world. The symptoms can be mild (an uncomfortable, burning sensation after eating, for example) or they can be much more severe. Ulcers (Figure B) can develop in the lining of the intestines, leading to pain, bleeding, and even perforation of the stomach or bowel, which is very serious. Left untreated, long-term inflammation increases the risk of cancer of the oesophagus or stomach. *H. pylori* infection plays a key role in the development of gastritis, but this was not recognised until the 1980s.

For many years, doctors thought most gastric ulcers were the result of stress, and the most common treatments were medicines to control acid production and surgery.

Figure A *Helicobacter pylori* under an electron microscope, clearly demonstrating its spiral shape.

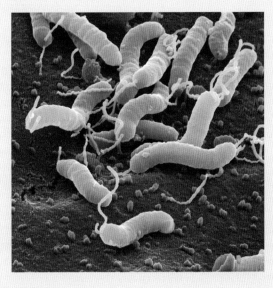

Juergen Berger/Science Photo Library

Figure B An ulcer in the wall of the stomach. Left untreated, this can lead to severe complications, and even stomach cancer.

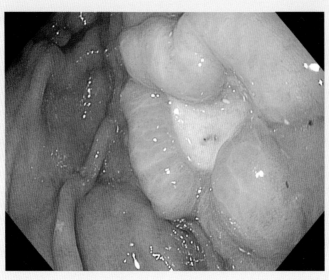

Gastrolab/Science Photo Library

In the late 1970s and early 80s, Barry Marshall and Robin Warren, two doctors at the Royal Perth Hospital in Australia, noticed that *H. pylori* bacteria were often present in biopsies taken from their patients with gastritis. In fact, inflamed stomach tissue seemed to contain the spiral bacterium more often than not—and it was much less common to find *H. pylori* in healthy stomachs. They suggested that the bacterium might be responsible for causing gastritis and its complications, but struggled to prove this using animal models. The scientific community remained sceptical, as definitive proof was difficult to obtain and many people had a vested interest in the existing treatments for gastritis—either drugs which needed to be taken for a lifetime, or large operations which required expensive, highly trained surgeons, theatre staff, and operating theatres. In 1982, Marshall decided to experiment on himself to prove the hypothesis. After undergoing an **endoscopy** to prove that he did not have any gastritis or pre-existing *H. pylori* infection, he drank the contents of a **Petri dish** containing the organism. He expected it to take some time for symptoms to develop, but within three days he began to experience abdominal discomfort and nausea, followed by reflux and vomiting. A further endoscopy performed two weeks after initial exposure to the bacterium demonstrated the presence of both gastritis and *H. pylori*. According to Koch's postulates (which you can find in Chapter 1, 'Koch's postulates of disease'), the link between infection and its associated condition (gastritis) had been proven. A relatively simple treatment regime including medication to suppress acid production and a course of antibiotics was enough to treat most affected patients, and would also help to protect them against gastric and oesophageal cancers in the future. In 2005 Marshall and Warren were awarded the Nobel Prize for their work—and their original research has stimulated other scientists to discover even more about *H. pylori* and its relationship with humanity.

Further research into *H. pylori* infection has revealed that our co-existence with this fascinating bacterium may be far more complicated than was initially thought. It is difficult to be certain how long it has been infecting humans, but it may well have happily existed in our stomachs for many thousands of years. With rates of gastric ulcer disease and gastric cancers falling over the past hundred years or so, but rates of gastric reflux disease and oesophageal cancers rising, there is potential evidence that *H. pylori* infection may be protective against some forms of cancer, even as it increases the risks of others. Some studies have demonstrated that children infected with *H. pylori* suffer from fewer cases of diarrhoea than children without it. Could there be a **symbiotic relationship** between humanity and *H. pylori*? Further research may yet reveal more to the story.

❓ Pause for thought

Think about the ethics involved in this case. Would it have been more ethical or less ethical for Marshall to infect others rather than himself? How might you be able to recruit volunteers for such a study in an ethical manner? How many people might you need to conclusively prove the link? Is a single case sufficient?

Look into the design of modern clinical trials and summarise the process. Compare this to Marshall and Warren's unorthodox approach and consider the pros and cons of each.

The structure and classification of bacteria

The basic structure of a prokaryotic cell is relatively simple, and something that you may already be familiar with. Look at Figure 2.3. It shows a diagram of a single bacterium, with several of the characteristics commonly found in different types of bacteria. Note that not every species of bacteria will possess all of the features in the diagram, and some bacteria (such as nitrogen-fixing bacteria in soil) will contain other specialised structures.

Figure 2.3 Generalised prokaryotic cell showing many of the features common to bacteria.

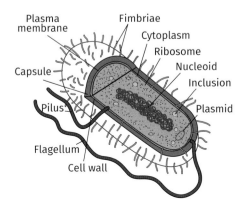

Many bacteria exhibit very similar structures, shapes, and organelles. They can share genetic code easily, so classifying them into distinct species can prove quite challenging for scientists. However, we can describe and differentiate bacteria in a number of ways. One simple way of classifying bacteria is by their shape. There are five common shapes, as shown in Figure 2.4.

These differently shaped bacteria can exist singly, in pairs, in clusters, or in chains. For example, *Neisseria gonorrhoeae*, the organism responsible for gonorrhoea, exists as a pair of bacteria. This is also referred to as a diplococcus. *Streptococcus pyogenes*, a bacterium responsible for several different types of infections in humans, including scarlet fever, forms a chain of cocci. *Staphylococcus aureus* forms clusters of cocci. It usually lives harmlessly on our skin, but can also be the cause of many different infections in humans. Some strains of *S. aureus*, such as methicillin-resistant Staphylococcus aureus (MRSA) have developed resistance to many different types of antibiotic, making them challenging to treat. *Streptobacillus moniliformis* is an example of bacilli which form chains. This bacterium causes a rare disease called rat bite fever.

In addition to these common shapes, bacteria can exhibit other unusual structural quirks. They may be surrounded by a thick, waxy coat (such as *Mycobacterium tuberculosis*, the bacterium responsible for causing tuberculosis). They may exist in organised clusters of four or more bacteria, or even geometric shapes such as cuboids.

Bacteria may also be classified depending on whether they require oxygen to survive, or whether they can exist in a low-oxygen environment. Those which require oxygen are called aerobes, and include *M. tuberculosis*. Those which can survive without oxygen are called anaerobes, and include *Escherichia coli* and *Klebsiella* species. Some bacteria can grow in both aerobic and anaerobic environments (such as many *Staphylococcus* species). They are known as facultative anaerobes.

Figure 2.4 Bacteria can be classified by their shapes.

Bacilli (rod-shaped bacteria), e.g. *B. anthracis*, an organism which causes anthrax in both humans and animals	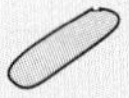
Cocci (spherical bacteria), e.g. *Micrococcus luteus*, which normally lives harmlessly on human skin, but can rarely cause opportunistic infections	
Spirilli (spiral bacteria), e.g. *H. pylori*, as mentioned in Case study 2.1, is a spiral-shaped bacterium	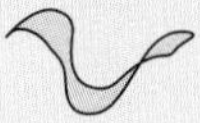
Vibrio (comma-shaped bacteria), e.g. *Vibrio cholerae*, the organism which causes cholera	
Spirochaetes (corkscrew-shaped bacteria), e.g. *Treponema pallidum*, the spirochaete responsible for causing syphilis, a sexually transmitted infection	

Bacteria may also be identified according to different staining techniques, which involve encouraging the bacterial cell wall to take up a coloured dye, making the bacterium much more easily seen under the microscope. Perhaps the most famous of these is the Gram stain, developed in 1884 by a Danish pathologist named Hans Christian Gram. He discovered that a dye named crystal violet was taken up by some bacteria but not by others. The bacteria which absorbed the dye were stained a bright violet colour when viewed under the microscope. Although he did not realise it at the time, Gram had struck upon a way of rapidly dividing many bacteria into two distinct groups. Gram-staining was useful in the identification of bacteria, but it had little practical use in medicine at the time it was developed—bacterial infections were still largely untreatable. During the twentieth century, however, it became a useful tool in the field of medical microbiology.

Quickly identifying whether a bacterium is Gram-positive or Gram-negative allows an appropriate antibiotic to be chosen long before the exact bacterium has been identified. In cases of serious infection (such as meningitis or septicaemia), delaying treatment by even a few minutes can mean the difference between life and death—so a rapid Gram-staining of samples will allow doctors to start treatment earlier than awaiting full culture results (which can take days). Gram's staining technique is still relevant and life-saving today, and many other stains exist to help identify different species of bacteria. Look at Figure 2.5—it shows the clear difference between Gram-positive and Gram-negative bacteria. You will learn more about Gram-staining in Chapter 3.

Further ways to identify and classify disease-causing bacteria, utilising advanced techniques such as genetic sequencing, will be discussed in more detail in Chapter 6.

Figure 2.5 Gram-positive organisms show up as bright violet when stained. Gram-negative organisms do not take up the violet stain, and appear pink instead. You will discover the reason for this in Chapter 3.

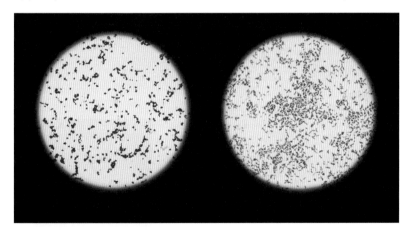

toeytoey/Shutterstock.com

Fungi—ancient, complex invaders

There are many thousands of species of fungus, but only a few of them cause disease in humans. They are ancient and remarkably complex, with very different structures to bacteria. Although individual fungal cells have a similar appearance to many other types of cell (with nuclei, organelles, and a cell membrane), the superstructure of fungi tends to be very different. With the exception of yeasts (which grow as individual cells), most fungi grow as thread-like filaments called hyphae. Each hypha consists of one or more fungal cells, surrounded by a tubular cell wall. The hyphae are divided up by septa—porous 'walls' which allow cytoplasm and organelles to flow through them. Look at Figure 2.6—this shows the structure of a fungal cell in greater detail, as well as the structure of a hypha. A network of connected hyphae is collectively called the mycelium. Some fungi are dimorphic—they are able to take on multiple forms throughout their lives. A good example is *Histoplasma capsulatum*, the fungus responsible for the disease histoplasmosis in humans. At temperatures of up to 25 degrees Celsius, this fungus exists as a mycelium. At human body temperature (around 37 degrees Celsius) it grows as single, round, yeast cells instead. Many fungi reproduce both asexually (through the formation of fungal spores) and sexually, using specialised organs to transfer reproductive material.

Our immune systems are quite well equipped to combat fungal infections, and our relatively high body temperature prevents many fungi from growing successfully in our bodies. Despite this, fungal skin infections are quite common—although they rarely cause any symptoms more severe than itching or redness. However, fungi are important pathogens for people with compromised immune systems—such as premature infants, transplant recipients, or those suffering from immunosuppression due to medication or HIV/AIDS. Diseases such as aspergillosis, histoplasmosis, or pneumocystitis pneumonia (PCP) can affect the lungs, causing symptoms of pneumonia such as breathlessness, cough, fever, and low oxygen levels. *Cryptococcus* species may infect the brain or its linings, causing symptoms of meningitis. Although rare, serious fungal infections are an important part of medical microbiology.

Figure 2.6 The structure of an individual fungal cell, extending into a hyphal tip. Note the pores in the septa, allowing exchange of contents between the cells. The second image shows the hyphae and fruiting bodies of the *Aspergillus* mould, which can cause severe lung infections for patients who have compromised immune systems.

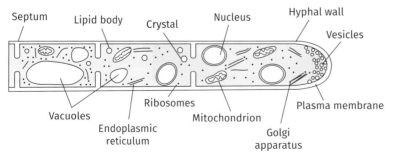

Right image: Dennis Kunkel Microscopy/Science Photo Library

Viruses—tiny replicators

Whilst bacteria tend to be relatively large and easy to see under the microscope, viruses are truly tiny—often measuring only a few nanometres in diameter. Because of their size, most of them are invisible under simple light microscopes, and can only be seen using electron microscopes. As a result, no one saw a virus until the second half of the twentieth century. Despite this, some nineteenth-century scientists theorised that these tiny infectious particles must exist, and a Dutch microbiologist named Martinus Beijerinck coined the term 'virus' to describe them.

Viruses are the ultimate parasites. They are usually made up of a strand of DNA or RNA (either double- or single-stranded), surrounded by a protein coat (called a capsid; Figure 2.7). Some viruses will also have an envelope comprised of a lipid bilayer (usually taken from the membrane of the host cell) studded with virus envelope glycoproteins. They have no cell membrane or wall, and no organelles at all, and because of this simplicity, they tend to be incredibly small, ranging from 20 to 400 nm in diameter. Viruses infiltrate the cells of their host, hijacking the host cell's organelles to transcribe their own genetic code and produce further viruses. They cannot reproduce without using the cells of their intended host, and are therefore said to be obligate parasites. Once the infected cell has exhausted its usefulness and is full to bursting with identical copies of

Figure 2.7 A basic viral particle or virion. A strand of genetic material (either DNA or RNA) surrounded by a capsid and an envelope. Viruses can change their protein coats extremely rapidly, which makes developing effective vaccinations or drugs very challenging.

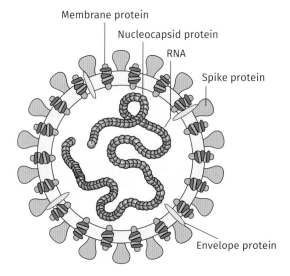

Adapted with permission from Rachel L. Graham, Eric F. Donaldson, and Ralph S. Baric, 'A Decade after SARS: Strategies for Controlling Emerging Coronaviruses'. Nature Reviews Microbiology. Volume 11, pages 36–848 (2013) https://doi.org/10.1038/nrmicro3143 Copyright © 2013, Springer Nature

the virus, it ruptures, releasing the viral clones, which will then go on to infect other cells in the host. Individual viruses have evolved to infect the cells of many different organisms, both animals and plants—with many viruses specialising in infecting only certain cell types in their hosts. Some viruses—called bacteriophages—have even evolved to only infect bacterial cells. This makes them very interesting to scientists looking at ways to reduce antibiotic resistance, and some potential use of bacteriophages in medical microbiology will be discussed further in Chapter 6.

Figure 2.8 Viral infiltration by fusing with the host cell membrane

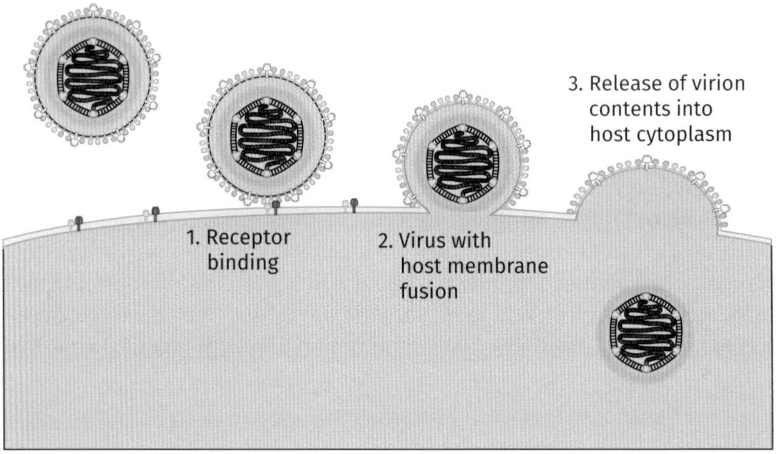

ViralZone, SIB Swiss Institute of Bioinformatics

Viral infiltration of the host cell occurs by one of three different mechanisms:

- Viruses without envelopes enter the host cell by a process called endocytosis. Glycoproteins on the surface of the virus's protein coat fuse with proteins on the host cell's membrane, and the virus is then absorbed into the cell, surrounded by a layer of the host's membrane—a vesicle. Once fully absorbed into the cell, the host membrane undergoes lysis, and the viral capsid then separates from the genetic code.

- Enveloped viruses may also enter using endocytosis. In this case, glycoproteins on the surface of the viral envelope fuse with the cell membrane, and the enveloped virus is again taken into the cell in a vesicle. Once fully absorbed into the cell, the envelope fuses with the vesicle and releases the capsid into the cell. The capsid separates from the genetic code.

- The final mechanism of entry used by enveloped viruses involves direct fusion of the envelope with the cell membrane, as you can see in Figure 2.8. Glycoprotein spikes on the viral envelope attach to proteins in the cell membrane, and the envelope then fuses directly with the membrane. The capsid and genetic code are then released into the cell, and the capsid subsequently separates from the genetic code.

Once inside the host cell, a virus has several challenges to overcome before it can successfully replicate itself. Firstly, it must hijack the cell's DNA translation and transcription organelles to read the viral genetic code. Secondly, it must prevent the cell from using its organelles as it normally does. Thirdly, it must perform the above actions without triggering the cell's own defence mechanisms, which would stop the organelles from functioning. All of these techniques must be encoded within the genetic code of the virus itself—which, given the tiny size of most viruses, is really rather amazing. Viruses have adapted many different ways of effecting these changes, suggesting that they must have evolved alongside their chosen hosts closely in order to counter any defences and successfully invade their target cells. Ultimately, if the virus is successful, the organelles within the cell will be forced into replicating viral genetic code and assembling new viruses. There is even evidence that some viruses will trigger changes within the cytoplasm of the cell, bringing appropriate organelles together to form 'virus factories' which allow for more efficient, rapid production of new viruses.

Once new viruses have been produced, they are then released from the cell, in a stage called viral shedding. This can occur in several distinct ways:

1. In enveloped viruses, the host cell membrane is often first studded with viral spike proteins. A viral matrix protein will then coat the underside of the membrane, and the capsid and genetic code of the virus will undergo 'budding', as it is covered by the spike-studded host cell membrane, forming the viral envelope. The virus is then released, ready to infect another host cell. This technique is primarily used by enveloped viruses, although it is also utilised by some non-enveloped viruses.

2. Some viruses are released using exocytosis. Whilst still within the cytoplasm, these viruses become encased in a vesicle which transports them through the cell membrane and releases them outside the cell. This technique is used by both enveloped and non-enveloped viruses.

3. The third mechanism of viral shedding is via programmed cell death, or apoptosis. Once the virus has replicated sufficiently, it will produce compounds which force its host to rupture, disrupting the cell membrane and spilling the contents of its cytoplasm, allowing the virus to spread, and killing the host cell. Both enveloped and non-enveloped viruses use this method of shedding.

4. Some viruses also take advantage of the body's own immune system. They will signal that the cell is infected, and once a macrophage (a cell type dedicated to absorbing infected cells and foreign bodies) engulfs the cell, the virus will then infect the macrophage as well. This method is primarily used by non-enveloped viruses, although some enveloped viruses (such as HIV) will use this technique to infect cells of the immune system.

Look at Figure 2.9 to see the life cycle of a typical viral infection.

Figure 2.9 Infection of an animal cell with a typical enveloped virus—influenza virus is a good example:

- Infiltration—the virus makes its way into the host cell, entering through the cell membrane. The viral capsid then releases the virus's genetic code into the cell's cytoplasm.
- Replication—the virus hijacks the cell's own organelles to replicate its own genetic code.
- Construction—the completed products from replication are assembled by the cell's own organelles, and prepared for the final stages. 'Virus factories' help to speed up this process.
- Shedding—once the host cell is full to bursting with identical copies of the virus, it is forced to shed the viruses, either by budding, exocytosis, or by total destruction of the host cell via apoptosis.

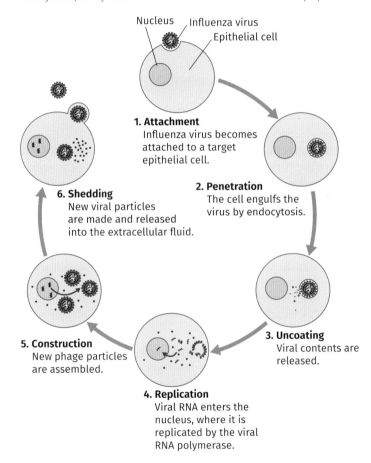

Nucleus   Influenza virus
          Epithelial cell

1. Attachment
Influenza virus becomes attached to a target epithelial cell.

2. Penetration
The cell engulfs the virus by endocytosis.

6. Shedding
New viral particles are made and released into the extracellular fluid.

3. Uncoating
Viral contents are released.

5. Construction
New phage particles are assembled.

4. Replication
Viral RNA enters the nucleus, where it is replicated by the viral RNA polymerase.

Classifying viruses

Viruses tend to be identified in one of several distinct ways. The International Committee for the Taxonomy of Viruses (ICTV) has suggested a scheme of 'classical' taxonomy, identifying viruses according to their order, family, genus, sub-genus, and species. Not everyone agrees with this way of identifying and classifying viruses, however. Look at the Baltimore classification in Figure 2.10. This aims to offer an alternative—grouping viruses

Figure 2.10 The Baltimore classification of viruses aims to group them according to their genetic similarities rather than placing them in family groups.

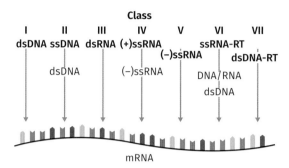

according to their genetic structure. This idea was originally suggested in the 1970s by a scientist called David Baltimore, as a more logical way of classifying viruses. By using their genetic material to categorise them, it becomes much easier to understand the relationships between different viruses. Classical taxonomy, on the other hand, does not immediately demonstrate these relationships.

Common viruses affecting humans (grouped using classical taxonomy) include:

- Rhinoviruses—Probably the most common virus to affect humanity; most of us will have contracted a rhinovirus at some point in our lives. They cause symptoms such as a sore throat, runny nose, dry cough, and fever—which, when combined, we term the common cold. They are tiny, non-enveloped viruses, often no more than 80 nanometres in diameter, comprising a single strand of RNA within a twenty-sided capsid (Figure 2.11). At least ninety different serotypes of rhinoviruses exist, and there is no successful treatment or vaccination available against them. They are spread via airborne droplets from coughs and sneezes.

- Rotaviruses—a genus of double-stranded RNA viruses which cause infectious diarrhoea. They are incredibly common, and it has been suggested that every child in the world will have at least one infection with a rotavirus by the age of five. They are spread by the faecal–oral route. Rotaviruses are small, non-enveloped viruses with icosahedral capsids, rarely exceeding 76 nanometres in size. Infections with rotaviruses kill over 200,000 people every year, especially in developing countries.

- Influenza viruses—viruses in the Orthomyxoviridae family, causing symptoms which we call 'flu'. Sufferers tend to experience high fevers, a dry cough, muscle aches, and can also experience diarrhoea and vomiting. Spread through airborne droplets from coughs and sneezes, outbreaks of flu tend to occur seasonally in the temperate areas of the world, but can occur at any time of year in the tropics.

Although many people recover completely from influenza, it still kills up to half a million people every year. Occasionally, pandemics of flu occur, which infect many millions of people. Influenza viruses are enveloped RNA viruses, between 80 and 120 nanometres in size. The RNA in influenza viruses is usually single-stranded, and exists in several distinct, segmented strands. Influenza viruses change their protein coats rapidly, but it is still possible to vaccinate against them, if the common strains for that particular outbreak year can be isolated. Influenza viruses are further classified by the proteins on their surface, grouping them into H and N groups. Have a look at The bigger picture 2.1 to learn more about identifying different flu viruses, and what happens when an outbreak becomes a pandemic.

There are, of course, many hundreds of other viruses affecting humans, causing anything from mild symptoms through to severe illness and death. The field of virology is constantly evolving, and we will look at treatments and vaccinations for viral diseases in greater detail in Chapter 6.

Figure 2.11 A rhinovirus. You have almost certainly been infected by one of these common viruses at some time in your life.

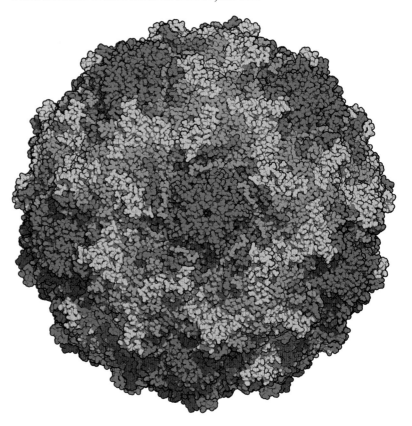

molekuul_be/Shutterstock.com

The bigger picture 2.1
Swine flu: a global pandemic

People in every country around the world suffer from flu. Countries with temperate climates tend to see a peak of cases in the winter, whereas those in the tropics may be affected all year round. Flu will often lead to high fevers, muscle aches, a cough, runny nose, and feeling generally very unwell. Patients will often be forced to rest in bed for a week or more, but usually make a full recovery. Sometimes it may make a sufferer more unwell, and if they are already in poor health, it can kill them. Although outbreaks of flu can be deadly to individuals, they rarely reach more threatening levels in the general population. The last great flu pandemic was in 1918, at the end of World War I, and it killed between twenty and fifty million people around the world. Unusually, many of the people were not old or very young and frail—they were fit young people in the prime of life. Almost one hundred years later, scientists began to see some worrying trends in a flu outbreak in central America . . .

In the early months of 2009, Mexican healthcare authorities noticed that an unusually large number of cases of influenza were being reported in the state of Veracruz. These cases did not seem to be limited to the elderly and those with long-term health conditions—the usual groups most affected by influenza. This outbreak affected all age groups, and appeared to cause unusually severe symptoms—it could lead to pneumonia, other breathing complications such as **respiratory distress syndrome**, and death, even in fit, healthy adults.

The outbreak spread within Mexico, and, before long, cases were being reported in the United States as well. Despite efforts by the Mexican Government to contain the spread of the flu virus (such as closing public and private facilities like swimming pools, schools, public transport, and museums), more cases were reported. The virus began to spread globally, with infected individuals continuing to travel internationally, carrying the virus inside their bodies even before they showed any symptoms, and spreading it via droplets when they coughed or sneezed. By late June, the **World Health Organization (WHO)** and Centre for Disease Control (CDC) had stopped counting new cases of the influenza and officially declared a global pandemic.

Through molecular analysis, the subtype of the influenza virus was identified as H1N1. The H and N letters refer to two proteins found on the surface envelope of the virus—haemagglutin and neuraminidase. They are used to classify all subtypes of the 'A' strains of the influenza virus. In April 2009, the CDC confirmed that this subtype of the influenza virus was new to humans, and appeared to be formed of a mixture of DNA from four different subtypes of flu virus. It appeared that the virus had initially appeared in pigs, which can also contract human influenza viruses. It was suggested that pigs had become infected with a mixture of different subtypes of the virus, and that the different viral subtypes had managed to share genetic material by infecting a cell at the same time. Research from later in 2009 showed that the H1N1 virus was capable of infiltrating much deeper into the lungs than typical seasonal influenza, which was why the new virus frequently caused more severe respiratory symptoms.

Countries around the world responded to the pandemic threat. People were warned about the virus, masks and antiviral hand gels were often used in public spaces, and anyone suffering from symptoms of influenza was encouraged to be tested (using throat or nasal swabs). Many people were offered antiviral medication such as oseltamivir (also known as Tamiflu), a medication which works to stop the action of the enzyme neuraminidase on the surface of the viral envelope. Neuraminidase helps the virus enter and exit its host cells, and inhibition by oseltamivir helps to prevent new copies of the virus from being released from infected cells. Tamiflu was also frequently offered to people who were felt to be significantly at risk from the more severe complications of H1N1 infection, such as those with weakened immune systems.

Eventually, the rate of infections began to drop. By May 2010, the number of cases was in steep decline, and in August the WHO declared that the pandemic was coming to an end. The aftermath led to a degree of confusion. According to the WHO's statistics, this pandemic had actually been significantly less deadly than the yearly seasonal flu outbreaks, with a mortality rate much lower than had initially been expected (although further analysis suggested that several hundred thousand deaths had not been attributed to the outbreak as they had not been confirmed in a microbiology laboratory). This led to critics claiming that the WHO had spread 'fear and confusion' during the outbreak, rather than information. It was suggested that some experts who had advised the WHO had links to pharmaceutical companies which produced either flu vaccinations or antivirals, and that the advice given to the WHO may not have been impartial. The WHO welcomed the criticism and conducted an internal review, after which they decided that the response to the pandemic had been appropriate.

❓ Pause for thought

You can read the WHO's archived information about the pandemic below. Use this to help you answer the following questions:
Link to CDC archive: https://www.cdc.gov/h1n1flu/cdcresponse.htm

- *How does an organisation like the WHO determine when an infectious disease has made the leap from localised outbreak to pandemic?*

- *How might this data be collected? What problems can you see in collecting data from around the world?*

- *One of the biggest criticisms against the WHO was that of 'fearmongering'. Do you think disease-monitoring agencies have an ethical duty to inform the public of outbreaks? If so, how should they work out when to inform the general public? Do they have a responsibility to do this in a sensitive fashion to attempt to reduce public panic, or is honesty the most important factor, no matter what the consequences are? Explain your opinions.*

- *The WHO was accused of being influenced by people with connections to pharmaceutical companies, who might have had an interest in seeing the dangers of this outbreak exaggerated. How might an organisation like the WHO identify and reduce bias from their advisers?*

Chapter summary

- Although there are many different types of pathogen, each group (eg bacteria, viruses, and fungi) tend to share similar traits which can be used to identify them.
- The success of bacteria is based on three features: rapid asexual reproduction by binary fission means beneficial genetic traits can be passed on rapidly; many bacteria can pass their genetic traits horizontally as well as vertically so beneficial traits can be spread rapidly within and between populations; and the ability to become super-specialised means they can take advantage of hostile environments.
- Bacteria may be grouped taxonomically, or according to shape, oxygen dependence, or staining.
- Many bacteria do not cause infectious diseases in humans. Those which do cause disease may lead to mild, localised infections or severe, life-threatening ones.
- Fungi are grouped taxonomically. They rarely cause infections in humans with functioning immune systems, but can take advantage of individuals with lowered immunity to cause severe symptoms of disease.
- Viruses have a variety of methods for infiltrating their host's cells, but cannot reproduce without using the organelles within these cells, making them obligate parasites. They will often damage or destroy the cell once they have replicated.
- Viruses tend to be classified either using classical taxonomic methods or by the Baltimore classification.

Further reading

1. **Niyaz Amed, '23 Years of the Discovery of *Helicobacter pylori*: Is the Debate Over?', *An Clin Microbiol Antimicrob* (2005). https://www.ncbi. nlm.nih.gov/pmc/articles/PMC1283743/**

   Gives a good overview of the possible evolutionary advantages of *H. pylori* infection.

2. **World Health Organization. http://www.who.int/csr/disease/swineflu/ frequently_asked_questions/about_disease/en/**

   WHO article on the H1N1 influenza virus and the global pandemic of 2009/2010, which also discusses reasons for the difference in expected and reported death rates.

3. **Centers for Disease Control and Prevention, 'Fungal Diseases'. https:// www.cdc.gov/fungal/index.html**

   The CDC's resources on fungal infections, including different types of fungal infections, discussing who is at risk, what types of fungi commonly cause infections in humans, and even listing recent outbreaks in the Americas and the Caribbean.

4. **Microbiology Online. http://microbiologyonline.org/**
   A website giving an overview of many different aspects of microbiology,
   including bacteria, viruses, and fungi. Run by the Microbiology Society,
   it is a wealth of information about microbes of all sorts, including those
   which cause human infectious diseases.

Discussion questions

2.1   Think about the discovery of a new pathogen (bacteria,
         virus, fungus, or parasite). How would you go about ethically
         demonstrating Koch's postulates to prove that this pathogen is
         the cause of a particular disease? How could you prove that the
         pathogen is different from any similar, existing examples?

2.2   Infectious diseases can be difficult to study in the laboratory, but it
         can also be challenging to study them inside the human body. What
         options do we have to further our knowledge of various pathogens,
         and what are the advantages and disadvantages of each one (e.g.
         animal models of disease versus studying human patients)?

2.3   Many infectious diseases have made the transition from an animal
         to a human host over time. Using modern scientific techniques,
         how could you prove the direction of transmission? Is there any
         way to work out how long this process might have taken?

A PANTHEON OF PATHOGENS 2: PARASITES, PRIONS, AND CULTURES

Like most other living things on planet Earth, humans can harbour many different species of parasite. Unlike bacteria and viruses, many of these parasites are visible with the naked eye (Figure 3.1)—they are often large, complicated, multicellular organisms, although some are tiny, single-celled creatures. Parasites have evolved alongside humanity, using our bodies as food supplies, homes, and nurseries. They lurk within our bloodstreams, on our skin, in our muscles, inside our intestines, and even within our nervous systems. Sometimes they cause no symptoms at all, and at other times they can exhaust our bodies to the point of death. Certain parasites have even evolved to manipulate human behaviour, changing the way we think or act to spread themselves more effectively. Some parasites only use us as a part of their life cycle. Others spend their entire lives inside their human hosts, only venturing outside our bodies to spread their offspring to others. When parasites infect us, they always have some effect on us—usually, but not always, causing an infectious disease.

Prions, another cause of infectious diseases, are even more mysterious, and have only been identified relatively recently. As you will discover, scientists are only just beginning to get to grips with these misshapen proteins, which can cause devastating diseases. This chapter will also examine different laboratory techniques for growing or identifying different types of pathogen—vital for diagnosis of disease and the selection of the appropriate treatment—as well as looking at the concept of **biosecurity**, the term used for a set of precautions that aim to prevent the introduction and spread of harmful organisms. As people live in ever bigger and more crowded cities, and as international travel becomes faster and cheaper, biosecurity becomes ever more important.

Figure 3.1 Not all organisms that cause infectious diseases in people are microscopic!

Photo: James Gathany. Public Health Image Library (PHIL)/CDC/Henry Bishop

Parasites—from single cells to giant worms

As you will learn in this section, parasites range from the microscopically tiny to the truly gigantic. Protozoa are some of the smaller parasitic organisms affecting humanity. They are single-celled creatures, often microscopic in size, which are able to move under their own power using cilia or footlike structures called pseudopodia. They either absorb nutrients through their cell membranes, or engulf their prey using phagocytosis. Many protozoa are entirely harmless to humans, but certain species have evolved to be efficient parasitic organisms, affecting different organ systems throughout our bodies. Probably the most well-known protozoan parasite is *Plasmodium*, the organism responsible for causing malaria in its human hosts. It has two stages to its life cycle, initially being carried in the salivary glands of mosquitoes before being injected into its human host. Once circulating within the host's bloodstream, it infects the liver before releasing further parasites into the bloodstream. These infect red blood cells and reproduce inside them (Figure 3.2), deforming them grotesquely and eventually

Figure 3.2 A red blood cell filled with plasmodia, the malaria parasite. The cell will eventually rupture, releasing more parasites into the bloodstream.

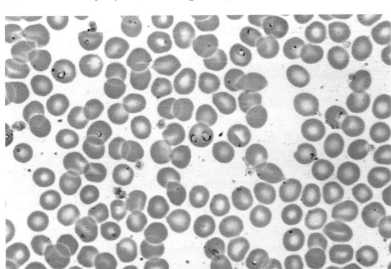

Public Health Image Library (PHIL)/CDC/Dr. Mae Melvin

rupturing them, flooding the bloodstream with new parasites and the contents of the red blood cell. It is the body's own immune response to the parasites in the blood stream (activating inflammatory pathways and releasing inflammatory chemicals called cytokines) which causes the symptoms of malaria—fever, rigors, chills, sickness and vomiting, diarrhoea, tiredness, aching joints, and even immunosuppression and death.

Other protozoal parasites invade other organ systems—*Giardia lamblia* (also known as Beaver fever) infects the intestines, causing diarrhoea, weight loss and abdominal cramps. *Trichomonas vaginalis* infects the vagina, causing itching, foul-smelling vaginal discharge, pain on urination, and increased risk of HIV infection. Leishmaniasis (caused by at least twenty different species of Leishmania trypanosomes, and spread by sandfly bites) causes ulcerations of the skin and mucous membranes, and may also lead to liver damage and immunosuppression.

Helminths are worms which can infest the human body. They are divided into three distinct groups—trematodes, cestodes, and nematodes. Most helminthic parasites live in our digestive tracts for at least part of their lives, although a few make more unusual homes elsewhere in the body—in blood vessels or individual organs. Almost all of them are visible with the naked eye in their adult stages. They all produce eggs (ova), which have evolved to survive in very hostile environments for prolonged periods of time, maximising their chances of surviving and finding a new host.

Flatworms

The trematodes are flatworms, ranging in size from just a few millimetres to up to 7 cm long. Commonly named flukes (a word taken from the Old English word for flounder and given to these worms for their flattened shape), they have relatively simple structures but surprisingly complicated life cycles. Many need two different hosts for the different stages of their reproductive lives. Often the first host is a mollusc, where the trematode reproduces asexually before moving on to its definitive host—which might be human or animal, depending on the species of trematode. Once inside the definitive host, the worm reproduces sexually, producing thousands of eggs which are released back into the environment, ready to begin the process again.

There are a number of different species of trematode which have chosen humanity as their definitive host, but perhaps the deadliest are the schistosomes (Figure 3.3). These are a group of flukes responsible for a disease known as schistosomiasis (also known as bilharzia or snail fever), classified by the WHO as the second most socioeconomically devastating parasitic disease worldwide. Passed from freshwater snails to its human hosts, schistosomiasis parasites enter through the skin, before passing into the blood vessels, where the flukes reproduce. They release thousands of eggs, which attempt to make their way to the bladder or intestines to be released into the outside world. The inflammatory reaction to these eggs causes bloody diarrhoea or bleeding from the bladder. Longer-term infection causes damage to the organs, as the eggs released by the parasite become engrained within the body's own tissues.

Figure 3.3 A male and female schistosome fluke; the female spends her life in a groove on the male's back, releasing thousands of eggs into the host's intestines and bladder.

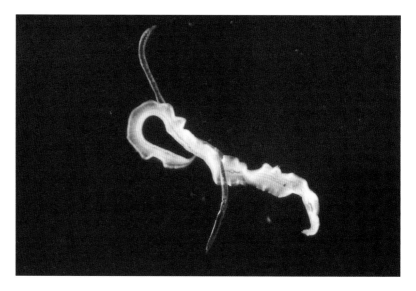

Public Health Image Library (PHIL)/CDC/Dr. D.S. Martin

Although the schistosomes themselves are adapted to avoid the immune system, any eggs which become encased in the body trigger a powerful immune response, causing inflammation and ultimately damage to the surrounding tissue. In long-term infection (which results in chronic, long-term inflammation) schistosomiasis can even lead to bladder cancers. If a child becomes infected, schistosomiasis can result in stunted growth and learning difficulties. In 2015, an estimated 250 million people worldwide were infected with the parasite, and it was implicated in many thousands of deaths. Because the life cycle of the parasite requires freshwater snails, it tends to affect areas of the world where clean water is in short supply, and is therefore most prevalent in the developing world. In areas where programmes have been introduced to eradicate freshwater snails and introduce clean water for drinking and bathing, schistosomiasis rates have been significantly reduced.

Tapeworms

Cestodes are the largest human parasites. This group of helminths includes the various different tapeworms which can infest our bodies. Each of these worms is formed of a long, multi-segmented body, attached to a headpiece called a scolex. Tapeworms undergo the early stages of their life cycle in an intermediate host—usually a herbivorous or omnivorous animal or a fish. Immature tapeworm larvae will encase themselves in the body tissue of the host animal, forming a cyst. They can remain dormant inside this cyst for long periods of time, but their ultimate goal is to be ingested (eaten) by their definitive host. Once the cyst has made its way through the digestive tract to the intestines, it releases the immature tapeworm. The scolex attaches itself to the wall of the intestines, using a system of hooks and suckers (which differs from species to species) (Figure 3.4). The immature tapeworm feeds on nutrients within the host's intestines, absorbing them through its skin, and it begins to grow. New segments of tapeworm—called proglottids—form, and the worm begins to increase in size. The proglottids furthest away from the

Figure 3.4 The scolex of a tapeworm, magnified many thousands of times. The hooks and suckers help it to attach to the wall of the intestine.

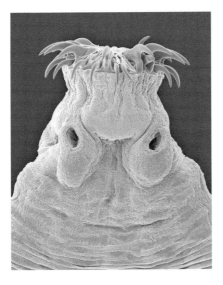

Dennis Kunkel Microscopy/Science Photo Library

scolex become mature, and are capable of fertilising and releasing eggs—tapeworms are hermaphroditic, and so do not need another tapeworm to reproduce. The eggs are released in the host's faeces, ready to be ingested by another host animal—either intermediate or definitive. The worm may live for a long time within the intestines, sometimes surviving for decades, and occasionally they reach a tremendous size (Figure 3.5). The longest tapeworm ever recorded in a human body was 33 metres in length!

Figure 3.5 If undetected, tapeworms can live for many years and attain a great length. Look at the photograph below. Can you work out how long this tapeworm is?

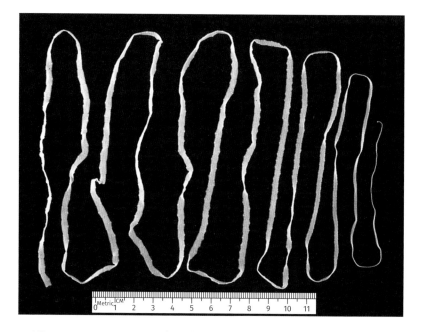

Public Health Image Library (PHIL)/CDC

Interestingly, despite their impressive credentials, most tapeworm infections don't cause any symptoms. Even if someone infected with a tapeworm does experience symptoms, they are often mild—upper abdominal discomfort, nausea, occasionally diarrhoea. With some species of fish tapeworm, the sufferer may become anaemic. Even more rarely, extremely large tapeworms may cause an obstruction in the bowels, requiring surgery. The real danger with tapeworm infestations occurs if a larval tapeworm finds itself in the wrong definitive host. When this happens, the larva's ability to migrate to the muscles of its host is compromised. Unable to navigate properly, it can form a cyst anywhere in the body. Depending on where the cysts form, this can lead to a variety of symptoms—and a disease called cysticercosis in humans. Ultimately, if any cysts form within the brain, they may cause seizures or headaches, and can be mistaken for brain tumours.

Usually, however, the only sign that a person has a tapeworm infestation is when they notice proglottid segments in their bowel motions. Once this has been observed, stool samples can be examined under the microscope for tapeworm ova, which confirms the diagnosis. Alternatively, blood samples can be tested for antibodies against tapeworm antigens, which may also give a positive result. The lack of symptoms is due to the tapeworm's ability to overcome our immune system. Like many other helminths, they are capable of 'damping down' some aspects of the immune response—reducing the number of T-helper cells and cytokines. This results in a less hostile environment for the worm to grow in, allowing it to exist relatively undisturbed inside its host.

Roundworms (nematodes)

The final group of helminthic parasites. Of the 25,000 different species of nematode, only a relatively small number have evolved to use humans as their hosts. However, these species vary tremendously—from the whip-worms, which embed their thin heads into the walls of the intestines, through to the long filarial worms which are the causes of filariasis and elephantiasis in the tropics (Figure 3.6). Some, like the roundworms and hookworms, migrate through the body as larvae, ending up in the lungs, where they are coughed up and swallowed into the intestines to complete the final stages of their life cycle. Others, such as the pin-worms, migrate to the surface of the anal skin at night to lay their eggs, causing itching around the anus and making it much more likely that the host will scratch and therefore transmit the eggs from anus to mouth, reinfecting themselves. Like the trematodes and the cestodes, nematodes are also thought to influence their hosts' immune systems, damping down the immune response and allowing the worms to reproduce and thrive. Different species of nematode may cause different symptoms. Some, like pin-worms, are harmless

Figure 3.6 Long-term infestation with filarial worms can lead to blockage of the lymphatic channels and enormous amounts of swelling in the limbs, genitals, or face.

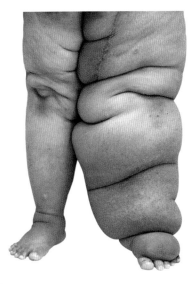

Tridsanu Thopet/Shutterstock.com

and may just cause some itching around the anus. Others, like hookworm or threadworm, may cause severe anaemia and bleeding from the intestines. Infection with some species may cause incredibly painful blisters as the worms reproduce under the skin, or hugely swollen limbs caused by blocked lymphatic channels.

The bigger picture 3.1
Parasites and protection—how some infections might help humanity

Infestation with parasites may seem very unpleasant, and the symptoms that they cause can make us very ill. People often find the idea of being infected with a parasite horrifying and disgusting—and many science fiction or horror films deal with just this thought. But perhaps parasitic infections aren't always as bad as they seem. Humanity has evolved alongside these creatures for millions of years, and evidence is emerging that perhaps our relationships with them are more complex than was originally thought. Could a dose of parasitic worms be good for your health?

In 2011, a geneticist and anthropologist named Kathleen Barnes and her team were studying the relationship between rates of asthma and one species of schistosomal parasite (*S. mansoni*, which you can see in Figure A) in a small cluster of fishing villages on the coast of Brazil. In this region, up to 85% of people are infected with the parasite, although some of them showed partial resistance to the infection. Dr Barnes and her team discovered that when they treated the villagers to rid them of worm infestations, rates of asthma and allergy symptoms within the community started to rise significantly. Infection with parasitic worms appeared to give some degree of protection against developing asthma or allergic symptoms. Other studies in Ethiopia, Ecuador, Gabon, and Venezuela appeared to support this theory.

Why might this be the case? What was the mechanism for the reduced rates of asthma, and why was this beneficial to the worms themselves? Several theories have been put forward. Firstly, many parasites can alter their hosts' immune responses, reducing the severity of the body's attempts to destroy them. Asthma and allergic reactions occur when the body's immune system responds inappropriately to something it perceives as a threat—pollen, dust, or other harmless particles which the body may come into contact with. If the entire immune system's aggressiveness is reduced by the parasitic infection, then a reduction in asthmatic or allergic symptoms makes sense. This gives the parasite an advantage—it makes the infected individual more likely to survive for long enough to procreate, thereby passing on the genes that make the host susceptible to parasitic infections. Dr Barnes and her team found that this appeared to be the case; people who were more resistant to schistosomal infestation were more likely to suffer from asthma.

Figure A *S. mansoni*, the schistosomal parasite studied by Barnes and her team.

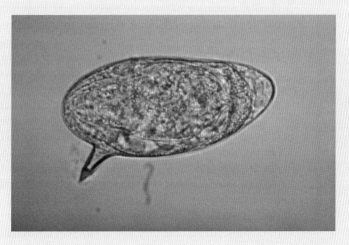

Public Health Image Library (PHIL)/CDC

Other theories have also been suggested to explain the relationship between parasites and allergic or immune system diseases. Some of the allergens specific to certain parasites have been shown to be chemically similar to allergens often detected in airborne particles, which can commonly cause asthmatic symptoms. Perhaps the immune system of infected individuals has already 'recognised' these allergens, and therefore does not respond to them as aggressively. A third theory has been called 'immune homeostasis'. This suggests that people with long-term parasitic infections have higher levels of inflammatory chemicals (such as cytokines and antibodies) circulating within their bodies. These chemicals signal to the body to start responding to threats, and it has been suggested that a higher level of these chemicals might raise the level at which the immune system starts actively responding, reducing the immune response accordingly but still leaving the individual able to respond to infections or other threats. Short-term reductions in the number of parasites within the body do not drastically affect the levels of inflammatory chemicals and antibodies, but longer-term reduction in the parasitic load can essentially 'reset' the body's sensitivities and trigger allergic or immune disease. All of these are currently theories, and it is likely that the true interplay between the human immune response and parasites is significantly more complex.

❓ Pause for thought

Although we may not yet fully understand the mechanisms, it is clear that parasites have the capacity to regulate our immune systems, with potentially very powerful effects. Think about the implications of this in the wider world. What effects might eliminating parasites in the developing world have? Would this be better or worse than the effects of parasitic diseases? Can you think of any ways we might be able to harness this effect, and use it to treat some medical conditions? How could this be safely and ethnically achieved and monitored?

Surface parasites

Finally, a number of parasitic species make their home on the surface of our bodies, living within our skin and hair. Headlice (*Pediculus humanus capitus*), body lice (*Pediculus humanus humanus*), and pubic lice (*Pthirus pubis*) are all insects (Figure 3.7). Unlike other ectoparasites, they spend their entire lives on their hosts' body, feeding on blood with their sharp mouth parts and moving from host to host via close body contact. If they are removed from the warmth and humidity of the human body, all species of human louse will quickly die—they are obligate parasites, and cannot survive without their host. Headlice and pubic lice lay their eggs on hair shafts, but the body louse lays its eggs in the seams of clothing. Headlice and body lice are very similar genetically, and it is thought that they only diverged as separate species about 10,000 years ago—which roughly coincides with the invention of clothes. The pubic louse is more distantly related to these other two species. Under laboratory conditions, headlice and body lice can still interbreed, but because they live in different environments on the host, they seldom (if ever) come into contact with one another. Headlice and pubic lice are harmless—they may cause some itching, but the lice themselves do not carry disease. Body lice, however, can carry bacteria such as *Rickettsia prowazekii*, the cause of typhus fever.

Figure 3.7 Although similar in shape and size, these three species of lice live in very different environments on the human body. Left: body louse; middle: head louse; right: pubic louse. All of them feed on blood.

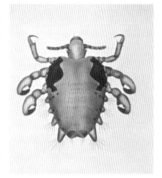

Left: Public Health Image Library (PHIL)/CDC; middle: Public Health Image Library (PHIL)/CDC/Herman Sander; right: Public Health Image Library (PHIL)/CDC/Dr. Dennis Juranek

Because lice are spread by close human contact, body lice can be important vectors for disease in susceptible populations—such as homeless people or crowded prison populations. Once detected (usually by looking at the lice with the naked eye or via a microscope), the lice may be eradicated using an insecticide such as permethrin or malathion. Headlice and their eggs (called nits) may also be combed out of the hair using a fine-toothed comb. The clothing of people infested with body lice should be washed at a high temperature to kill any lice lurking in the clothes themselves, and to prevent any eggs from hatching. Clothing should be changed at least once weekly, and regularly washed, to reduce the risk of recurrence of the lice. Other ectoparasites include the scabies mite *Sarcoptes scabiei*, which burrows under the skin and causes intense itching, as the host's immune system reacts to the faeces from the mites.

The human botfly is one of several fly species to parasitise humans, using the upper layers of their skin as a nursery for its offspring. The botfly will attach its egg to one of several different species of bloodsucking mosquito. When the mosquito lands on a person and begins to feed, the egg will drop off and quickly hatch, with the larva burrowing into the skin. Once settled, the larva grows and develops over approximately eight weeks, before emerging and dropping into the soil to pupate (Figure 3.8). Botfly maggots cause pain and swelling, and may cause other symptoms if they burrow into sensitive tissue such as the eyelids. They do not commonly cause bacterial infections, however, and it has been suggested that the maggot possibly secretes antimicrobial compounds to prevent its home from becoming infected. Treatment of the infestation involves either suffocating the maggot (using petroleum jelly or glue smeared over its breathing hole) and removing it with tweezers, or the use of oral ivermectin, an insecticide, which forces the maggot to exit its burrow and drop off. Other creatures, such as ticks and fleas, may temporarily infest humans to feed on our blood. They can spread diseases such as bubonic plague and Lyme disease, passing them on when they bite their host.

Figure 3.8 Infestation with a botfly maggot may cause pain and swelling over the affected area. Left undisturbed, the maggot will grow inside its host for eight weeks before dropping out to pupate.

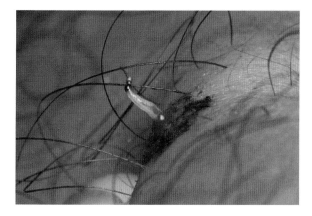

Amazon-Images/Alamy Stock Photo

Prions—misshapen killers

All the infectious diseases we have studied so far, both in Chapter 2 and this chapter, are caused by relatively complex organisms. From the smallest virus to the largest tapeworms, all of them contain some form of nucleic acid, either DNA or RNA. However, some of the rarest and deadliest infectious diseases are caused by single, misfolded proteins called prions (a combination of the words protein and infection). We still don't fully understand these tiny killers, but we know that they are capable of entering healthy hosts and converting other proteins into their misfolded form. The altered proteins cannot fulfil their normal function within the body, and eventually, as more and more proteins are affected, the host begins to suffer from symptoms.

The protein initially identified as being responsible for prion diseases was named PrP by its discoverers. This protein is found in the body of healthy humans and animals, where it occurs naturally on cell membranes. The normally structured protein is named PrP^c, with the 'C' standing for 'cellular'. The misshapen form is named PrP^{sc}—the 'SC' standing for 'scrapie', the first prion disease ever identified. The function of PrP in healthy individuals is not yet understood, and animal studies have provided only limited data. In 2015, another protein was implicated in the disease multi-systems atrophy (MSA)—the first prion disease thought not to be caused by PrP. All of the other known prion diseases—including Creutzfeldt–Jakob disease (CJD), fatal familial insomnia (FFI), and Gerstmann–Sträussler–Scheinker disease—are caused by faulty versions of PrP.

Prion diseases may be inherited, acquired, or may occur sporadically (through misfolding of normal PrP in the host). Once the prion protein enters a host, it induces misfolding in the normal prion proteins. These misshapen proteins begin to group together, forming clumps around tissue in the central nervous system. These clumps are known as amyloid plaques, and they begin to disrupt the normal tissue structure. Ordinarily, proteins are broken down by enzymes within the body when they have fulfilled their purpose. Prion proteins are resistant to this process, and so they continue to accumulate, causing progressive symptoms. Because the proteins build up mainly in the nervous system, they cause symptoms such as dementia, personality change, insomnia, and abnormal movements. There is no known cure for any of the prion diseases, nor are there any effective treatments. All prion diseases are degenerative, progressive, and fatal.

Some evidence has recently emerged that diseases such as Huntington's chorea (a genetic disorder, in which incorrectly coded proteins build up in the brain, causing dementia and movement disorders) could possibly be treated using artificial DNA. Researchers in London discovered that, after they injected Huntington's sufferers with synthetic DNA molecules, they were able to significantly slow down the rate of production of the Huntingdon's disease protein. This synthetic DNA altered the function of messenger RNA, stopping it from translating the genetic coding for the faulty Huntington's protein (known as huntingtin). This reduced production of huntingtin. This was a very small clinical trial, involving only forty-six patients, but the implications are tremendous. If this method of synthetic DNA therapy provides a successful cure for Huntington's disease, it might be possible to treat other neurodegenerative diseases caused by abnormal protein build-ups, such as Alzheimer's, Parkinson's—or prion disease.

Case study 3.1
Kuru—the laughing death

Prion diseases are still not well understood—we cannot predict or cure them, and even diagnosis often takes many years, as the symptoms may mimic those of many other different illnesses. However, we at least have the advantage of modern medical knowledge to guide us through the diagnostic process. To people without that knowledge, prion diseases must have seemed even more terrifying and incomprehensible.

The tribespeople of the highlands in Papua New Guinea were not discovered by the rest of the world until the early twentieth century, when Australian gold prospectors first described them. In the 1950s, researchers arriving to study the culture and habits of these people made a disturbing discovery. In a tribe called the Fore (Figure A), nearly 200 people were dying every year from a disease that the tribespeople called Kuru, meaning 'trembling' or 'shivering'. The disease usually killed its victims within twelve months—beginning with difficulties walking, but quickly progressing to a complete loss of control over the limbs. Victims would lose control over their emotions, laughing or crying hysterically, and became incontinent, unable to feed themselves or move from the floor. Once a victim had reached this stage, they would quickly die. The victims were almost all women, or children under the age of eight— older children and adult men were mostly spared from the grisly fate. The Fore tribe was only made up of about 11,000 individuals, and Kuru had killed

Figure A The Fore tribespeople of Papua New Guinea suffered from a terrible degenerative disease caused by prions.

Wellcome Collection

enough women for the tribespeople to be at risk of dying out—in some villages, there were barely any young women left.

The Fore people believed that the disease was a result of evil sorcery, and despite ruling out huge numbers of environmental causes, researchers were baffled as to the true cause. It clearly wasn't a bacterium, and it didn't appear to be a virus, or a parasite. It didn't seem to be caused by anything easily identifiable in the surroundings. Could it be inherited, the result of faulty genes?

Shirley Lindenbaum, a **medical anthropologist**, travelled from village to village to obtain family trees, and realised that the disease couldn't be genetic—it affected women and children in the same social groups, but not always in the same families. It also became clear that it had started in the North of the country, and had migrated southwards. Lindenbaum wondered if the disease might have its origin in one of the tribal customs surrounding funerals, for the Fore people were one of a small number of civilisations who still practised cannibalism. Once a person died, their body would be cooked and eaten by the close friends and family of the deceased. This was an act of great respect—the tribespeople believed that this was a way of helping the spirit of the dead person, which also prevented their physical form from being eaten by worms or maggots. The Fore people also believed that only an adult woman could control the powerful spirit of the dead person, and so they were the primary consumers of the body parts. Children young enough to still be living with the women would sometimes be given scraps to snack on. Almost every part of the body was eaten—including the brain and spinal tissue.

Lindenbaum turned out to be entirely correct. Teams of biologists flew out to study the disease, and at the time dubbed the condition 'a slow virus'—winning a Nobel prize for their work. It would not be until many years later that the true cause of Kuru would be discovered—prions. Even in remote Papua New Guinea, these tiny misshapen proteins had made themselves felt. As for the original source of the prion disease, it seems likely that this was from a case of sporadic CJD—a spontaneous misfolding of prion protein in a single individual. This misfolded protein would have slowly replicated itself within its host, causing the symptoms of the first case of Kuru. When they died of the disease, the grieving family would have consumed the body, unwittingly infecting themselves and spreading the disease. Although cooking the body parts would have killed most bacteria or viruses, prion proteins are heat resistant and can survive the cooking process. From a single source, the prion disease had spread throughout the tribespeople of the Fore, slowly spreading until it threatened the existence of an entire tribe of people.

Once the cause had been identified and the Fore people educated as to the probable cause of the sickness, the mortuary feasts came to an end. The cases of Kuru did not stop straight away, however. Because prions can remain in the body for years or even decades, slowly replicating until they produce symptoms, many Fore people developed Kuru years after the tribe had stopped practising cannibalism. The last confirmed death from Kuru was in 2009, and the medical team monitoring the Fore people declared the outbreak officially over in 2012. Since then, there have been rumours of other cases of Kuru, but further investigation has revealed no confirmed diagnoses. For now, Kuru is an extinct disease.

❓ Pause for thought

Think about the challenges facing scientists when they are confronted with a new disease. How might you go about eliminating possible causative factors? What patterns might be helpful in identifying the cause of the disease and the mode of its transmission? What other challenges might arise if the disease turns out to be caused or spread by cultural traditions? How might you address those challenges?

Growing pathogens in the laboratory

In order to successfully study pathogens, scientists need to be able to grow them in a laboratory environment, measure the phases of their growth, and visualise them using microscopy or other techniques. Potentially dangerous pathogens must be grown in a secure environment, to reduce the risks of infecting the scientists working with them or releasing dangerous infectious diseases into the wider world.

Culturing bacteria and fungi

Many species of bacteria and fungi can be grown in a laboratory environment, although some are easier to culture than others. Many viruses are very difficult to culture. You may have some experience of culturing bacteria in Petri dishes using agar jelly. The procedure for culturing organisms in a research or medical laboratory environment is similar—although different culture media may be used, depending on the circumstances. The most commonly used different types of media include:

- Nutrient medium—a source of proteins (amino acids) and nitrogen—often derived from beef extract or yeast extract. This medium is undefined, as the exact composition of amino acids cannot be determined. Containing all of the elements that many bacteria and fungi need to survive, nutrient media is often used for general cultivation and maintenance of cultures in laboratories.
- Minimal media—containing the bare minimum of nutrients that microorganisms need to form a colony, often without the presence of selected amino acids. This type of medium is often used to study 'wild' type organisms—but to culture them successfully, their requirements must be known in some detail.
- Selective media—used to specifically culture a certain type of microorganism (e.g., see Figure 3.9). For example, if the microorganism to be cultured is resistant to a certain type of antibiotic, the medium may be impregnated with the antibiotic, which will prevent any other species from growing.

- Differential media—also known as indicator media, these are used to distinguish one type of microorganism from another. Using the biochemical characteristics of an organism, certain nutrients or chemicals may be added to the medium to induce a visible indication which may define the microorganism in question—for example, blood agar can be used to determine which species of *Streptococcus* bacteria are present in a sample.

- Transport media—used in the transport of specimens, these are temporary media designed to halt replication, but not kill the microorganisms in transit (see Figure 3.10 for an example). They contain no amino acids, only buffers and salts, and may have other requirements depending on the microorganisms—for example, transport media for anaerobic bacteria must not contain any molecular oxygen.

Figure 3.9 Selective media for growing gonorrhoea. The plate on the left shows considerable bacterial overgrowth (white lines), whereas the selective media on the right shows only gonorrhoea colonies (white spots).

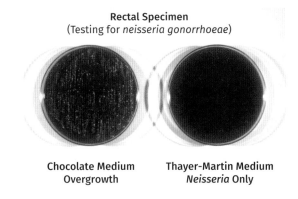

Rectal Specimen
(Testing for *neisseria gonorrhoeae*)

Chocolate Medium   **Thayer-Martin Medium**
Overgrowth   *Neisseria* Only

Public Health Image Library (PHIL)/CDC/Renelle Woodall

Culture media may be solid, semi-solid, or liquid (broth), depending on the amount of gelling agent added to them. This may be agar or gelatine, with different amounts of gelling agent being required depending on the purpose of the culture medium.

Figure 3.10 Bacterial swab, used to take samples of liquid material (e.g., pus from a wound) and transport it to the laboratory.

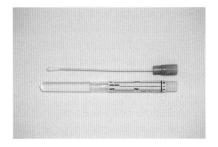

zairiazmal/Shutterstock.com

Bacterial growth patterns

Once deposited into the correct media, bacteria undergo several different phases of growth (Figure 3.11):

- Lag phase, during which the bacteria are not ready to divide—they are adapting to the new medium, maturing, and preparing for division.

- Log phase—here the bacteria begin to divide, and the phase is characterised by logarithmic cell division—the cells divide exponentially, and their population increases quickly. However, in a culture medium, resources will soon be exhausted by this process, and the bacteria enter the third phase.

- Stationary phase—caused by exhaustion of a factor key to bacterial growth, such as the depletion of an essential nutrient. At this stage, growth rate and death rate are equal, and the population of bacteria remains static.

- Decline phase, during which the population of the bacteria begins to fall. This may be because of lack of nutrients, environmental changes (e.g., a rise or fall in temperature), or other conditions, such as the introduction of antibiotics to the culture. During the decline phase, bacterial population usually decreases exponentially—essentially the reverse of the log phase.

Figure 3.11 Phases of bacterial growth.

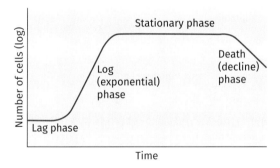

Once the microorganism has been successfully cultured, it may be examined under the microscope. Sometimes direct microscopy will allow adequate identification of the microorganism concerned. At other times, the organism may need to be stained to better identify it. This involves washing a microscope slide with dye, which is taken up by the cells of the microorganism to make them more easily identifiable under the microscope. There are many different stains, which are used to identify many different types of microorganism, but perhaps the most famous is the Gram stain, discussed in Chapter 2. You can see the technique for Gram-staining in Scientific approach 3.1.

Scientific approach 3.1
Gram-staining technique

As you have already seen, Gram-staining is one way in which scientists can quickly differentiate between different types of bacteria. Using a particular dyeing technique, bacteria can rapidly be identified as 'Gram-positive' or 'Gram-negative', potentially allowing a patient to be started on effective anti-biotics before the identity of the bacterium is fully known.

The reason that Gram-staining works is due to the composition of bacterial cell walls. Gram-positive bacteria contain a thick layer of a protein called peptidoglycan, embedded within their cell walls. Gram-negative bacteria have a much thinner layer of this protein, which tends to be much deeper within the cell wall. You can see a diagram demonstrating this peptidoglycan layer in Figure A.

Figure A These diagrams show the composition of cell walls in both Gram-positive and Gram-negative bacteria. The green layer in each is composed of peptidoglycans. You can clearly see that the peptidoglycan layer (the green molecules) in the Gram-positive bacterium is much thicker and closer to the surface than in the Gram-negative bacterium.

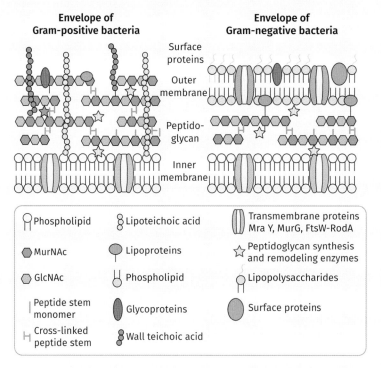

The Gram-staining process involves using two separate dyes—crystal violet and safranin. In the initial stages, the microscope slides are prepared by washing them with soap and water, before wiping them down with ethyl alcohol. This prevents the slides becoming contaminated with dust, dirt, grease, or unwanted microorganisms. A smear of bacteria from the desired culture or agar plate is then taken and placed on the slide, using a sterilised inoculation loop (which you can see in Figure B). At this point, the slide is then passed through the flame of a bunsen burner, 'fixing' the culture. This kills the bacteria and ensures that the culture adheres to the slide.

The next stage involves applying the first stain. The slide is gently flooded with crystal violet, and is then left to stand for one minute. Crystal violet is taken up by the peptidoglycan layers of both Gram-positive and Gram-negative cells, making them both appear a deep purple colour under the microscope at this point. The next step is to flood the slide again, first with Gram's iodine solution, and then with ethyl alcohol. The iodine 'fixes' the crystal violet stain to the peptidoglycans, and the ethyl alcohol then destroys the thinner layer of peptidoglycans present in the Gram-negative cell wall. If you were to look at the slide under the microscope at this stage, you would see that any Gram-positive bacteria would clearly show up as deep purple, but that Gram-negative bacteria wouldn't be very visible at all. The final stage of the staining process is to counter-stain the slide with safranin. This is taken up by the Gram-negative bacteria, making them appear pink, in comparison to the deep purple of the Gram-positives. Finally, the slide is flushed with water and blotted dry, before being examined under the microscope in order to work out whether the bacteria from the culture are Gram-positive or Gram-negative.

Figure B This series of diagrams outlines each step in the Gram-staining process.

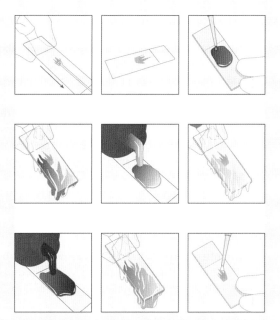

Peter Gardiner/Science Photo Library

However, not every species of bacterium can be classified as Gram-positive or Gram-negative. Some of them do not take up either of these stains easily, due to differences in the structure of their cell walls. In these cases, a different form of staining may have to be used. Many other stains exist, each of which has been designed to help identify a specific type or group of bacteria (see Figure 3.12 for examples).

Figure 3.12 Acid fast staining (left)—used to identify *Mycobacterium* species—and India ink staining (right)—used to identify cryptococcal fungi.

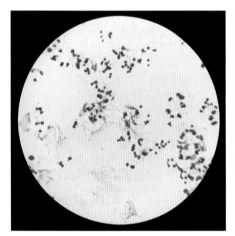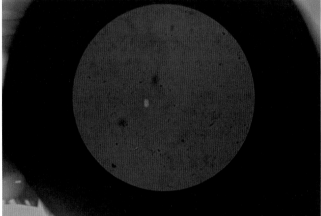

Left: Sirikwan Dokuta/Shutterstock.com; right: Chadsikan Tawanthaisong/Shutterstock.com

Parasites and viruses

Parasites cannot usually be grown under laboratory conditions, as they require a living host to feed and reproduce. Because many of them are quite large, they can often be directly visualised with the naked eye, and may be first seen when they are passed out of the host's body—infected people may notice the proglottids of a tapeworm in their stools, or a botfly maggot emerging from their skin. However, it is also often possible to look for the microscopic larval stages or the eggs of a parasite in samples sent to a laboratory (Figure 3.13). Often these are stool samples, as many parasitic species make their homes inside the intestines, but other samples such as skin biopsies may also be used for different parasitic species. The biochemical and immunological tests discussed below for diagnosing viral infections may also be useful in the diagnosis of parasites, if direct microscopy is inconclusive.

It is often very challenging to grow viruses in the laboratory, and many other parasitic species including various protozoa and worms are also very difficult to keep alive once they are removed from their hosts. A different approach must be taken to study these obligate parasites in the lab, and several techniques are used to do so. Viruses cannot be seen with the naked

Figure 3.13 Liver fluke ova seen under the microscope. The presence of the ova is one way in which infection with these parasites can be diagnosed.

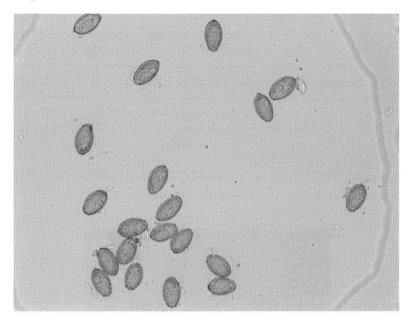

Peerachart/Shutterstock.com

eye, but may be visualised inside infected cells using electron microscopy. Samples of infected tissue may therefore be examined under an electron microscope to look for physical evidence of viral infection. This process is slow and not very reliable, as an observer may have to examine many different cells before they see any evidence of viruses. A more effective way to look for evidence of viral infection is to look for the biochemical markers left behind by viruses themselves, or by the immune system's response to infection. Samples of tissue from the infected person can be examined for either antigens or antibodies using immunoassay techniques, which are now mostly automated, and can be accomplished very quickly. This allows scientists to look for many different types of virus with a single sample, and gain a result far more quickly than would be possible using just electron microscopy. Scientists can also look directly for the genetic code of the virus, using a technique called polymerase chain reaction (PCR). This technique reproduces a tiny section of genetic code many thousands of times to make it easier to detect, allowing very small amounts of sample to be used whilst still obtaining an accurate result. It has proved extremely useful in identifying infections which are normally very hard to detect, such as HIV. Occasionally it may be possible to directly visualise the effects of some of the larger viruses within infected cells. Ordinary light microscopy may be used for this, and the changes within the cells often reflect the individual virus type affecting them—although this method is obviously less reliable than many of the ones described above.

Biosecurity—reducing the risk of transmission

Working in laboratories that regularly handle samples contaminated with infectious diseases requires certain precautions (for example, Figure 3.14). It is obviously hugely important that individual scientists are protected from accidental infection with any diseases, and equally important that diseases are not released from the lab, where they might be able to infect people in the general population. The steps taken to manage these risks are given the name biosecurity, and will vary significantly depending on the types of infectious disease that the laboratory deals with. The World Health Organization (WHO) classifies infectious agents in four different groups depending on their virulence, ease of transmission, and availability of treatment, as you can see in Table 3.1.

Figure 3.14 Scientists may use positive-pressure suits whilst working with particularly hazardous pathogens.

CDC/Alamy Stock Photo

When these measures are successfully followed, the risk of infection to individual workers or the general population is very low. However, if biosafety measures are inadequate for the pathogen being handled it can result in outbreaks of extremely deadly diseases. Look at Case study 3.2, which shows what can happen when biosecurity measures fail.

Table 3.1 WHO classifies pathogens into four groups, and recommends that laboratories handling more dangerous organisms adhere to stricter biosecurity measures.

	Biosafety level			
	1	2	3	4
Isolation[a] of laboratory	No	No	Yes	Yes
Room sealable for decontamination	No	No	Yes	Yes
Ventilation				
– inward airflow	No	Desirable	Yes	Yes
– controlled ventilating system	No	Desirable	Yes	Yes
– HEPA-filtered air exhaust	No	No	Yes/No[b]	Yes
Double-door entry	No	No	Yes	Yes
Airlock	No	No	No	Yes
Airlock with shower	No	No	No	Yes
Anteroom	No	No	Yes	–
Anteroom with shower	No	No	Yes/No[c]	No
Effluent treatment	No	No	Yes/No[c]	Yes
Autoclave:				
– on site	No	Desirable	Yes	Yes
– in laboratory room	No	No	Desirable	Yes
– double-ended	No	No	Desirable	Yes
Biological safety cabinets	No	Desirable	Yes	Yes
Personnel safety monitoring capability[d]	No	No	Desirable	Yes

[a] Environmental and functional isolation from general traffic.
[b] Dependent on location of exhaust (see Chapter 4).
[c] Dependent on agent(s) used in the laboratory.
[d] E.g., window, closed-circuit television, two-way communication.

Case study 3.2
A breach in biosecurity

In 1967, scientists working at laboratories in Marburg, Frankfurt, and Belgrade started to fall ill with very similar symptoms. It didn't seem too serious at first—fevers, headaches, muscle aches, and a general feeling of malaise. Some of them took time off sick, assuming that they had picked up a virus—

perhaps a dose of the flu. Cared for by their family members, they stayed in bed, rested, and waited to recover. By the end of the first week, more alarming symptoms began to appear. Victims started to experience diarrhoea, extreme nausea, and vomiting. Some were admitted to hospital amidst concerns about typhoid or dysentery. But this disease was not typhoid or dysentery, and sufferers continued to become more and more unwell. During the second week of illness their temperatures fell, but people's bodies began to struggle against the infection overwhelming them. They began to bruise easily, then to haemorrhage from every orifice. Even needle marks from taking blood samples resulted in significant blood loss. Their red and white blood cell levels fell catastrophically. Their livers, and then many of their other organs, began to fail. Some of the scientists died from severe haemorrhagic shock, their bodies simply unable to cope with the amount of blood that they were losing. Others lingered on in hospital as the infection ran rampant through all of their organ systems. Many became confused and had altered sensation in their limbs—signs that the infection had reached their brains. Those who did not succumb to the infection slowly began to improve. Eventually they would manage to fight off the infection—but many of the sufferers would go on to have painful, dangerous relapses for many weeks, with uveitis (inflammation in the eyes), hepatitis (liver inflammation and damage), and orchitis (inflammation of the testes). In total, between the three laboratories, thirty-one people had shown symptoms. Of these people, seven had died, and the health authorities had no idea what had caused the outbreak.

An investigation began to identify both the new disease and its source. The disease had seemed similar to some of the haemorrhagic fevers known about at the time, but blood tests for antibodies to these were all negative. Whilst the victims were still struggling for life in hospital, samples of their blood were flown to microbiology labs both nationally and internationally to try to identify the cause and prevent any more deaths. Studies in animal models were initially inconclusive. Eventually however, the disease was replicated in guinea pigs. Inoculated with blood samples from infected patients, they too became sick with similar symptoms, and quickly died. Neither conventional nor electron microscopy had helped to identify any obvious pathogen, but a technique called immunofluorescence (which uses fluorescent dyes attached to antibodies produced by the body) demonstrated some unusual appearances in the cytoplasm of some infected cells. Further electron microscopy eventually identified the culprit of the outbreak—a tiny, thread-like virus which had never been seen before (Figure A). Dubbed Marburg virus for the location of the first outbreak, it was the first in the filovirus family, which now also contains the Ebola viruses (see Figure A).

The virus had been identified, but where had it come from? Clearly something in the laboratories was to blame, and it rapidly became apparent that each of the labs had received two shipments of Green African Monkeys shortly before the outbreak. Contact tracing revealed that all of the victims had either had direct contact with tissue from these monkeys, or were closely related to somebody who had. The virus didn't seem to be airborne—not everybody in the laboratories had become infected, and not everyone who had experienced close contact with sufferers had contracted the disease. Further investigation

Figure A A scanning electron microscope image of the virus responsible for the outbreak.

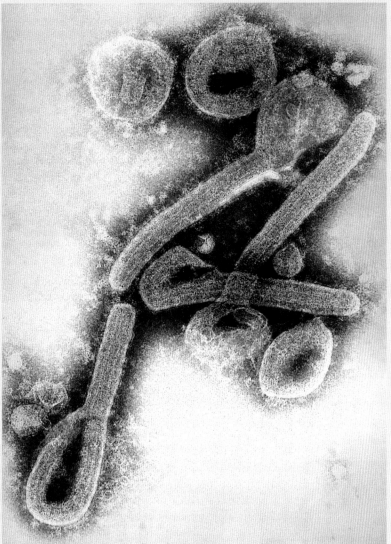

CDC/Fred Murphy/BSIP/Age Fotostock

revealed that the virus was spread through bodily fluids, or exposure to infected tissue. It seemed that everybody who had been exposed in this way had contracted the disease—everyone in the laboratories who had remained symptom-free went on to test negative for the virus—and the monkeys had seemed in good health when they arrived at the laboratories. The incubation period for Marburg virus was short—only a few days between exposure to the virus and the first symptoms developing—and the monkeys had been in transit for several weeks, more than long enough for them to start showing signs of infection. So how had the primates carried the disease?

The answer to this question was a mystery for years. New evidence suggests that it may be partly due to the techniques of the hunters who originally caught the monkeys. It appears that if a monkey caught by the hunters did not seem to be healthy, they would deposit the monkey on an island in a lake, where it would either die or recover. If the hunters couldn't catch enough monkeys in the wild, they would make their way over to this island and capture one or two of the healthier-looking monkeys that they had marooned there earlier. We now know that Marburg virus can lie dormant in reservoirs such as the testes or the eyes for many months, even after someone has fought off the disease. It therefore seems likely that the monkeys infected with the virus had survived the infection, but were still carrying copies of the virus in their bodies. The Marburg outbreak in Germany settled, and the virus has only occurred sporadically since then, although larger-scale outbreaks in the Democratic Republic of Congo (DRC) and Angola have claimed several hundred lives since 1998.

❓ Pause for thought

Think about the biosecurity techniques which are necessary in a microbiology laboratory. Using the WHO guidelines in Table 3.1 to help you, think about how you might try to prevent a failure of biosecurity like the one which led to the Marburg outbreak. What particular challenges did the investigators in the Marburg outbreak face? How might you be able to reduce these challenges in the case of a similar disaster in the future?

Of course, biosecurity doesn't just apply to laboratories. At a very basic level, even washing your hands after using the toilet is a biosecurity measure—you are reducing your risk of transferring potentially harmful bacteria from your hands to your mouth. Covering your mouth and nose when coughing or sneezing is another, potentially very effective, biosecurity measure. These simple measures can often have significant impacts, and not just on health. The WHO found that when handwashing techniques were introduced to schools in China, for example, the rate of absences amongst students fell significantly—improving the children's access to education. Handwashing is considered so important in health care across the globe that the WHO publishes a list of 'five moments for hand hygiene' for use in hospitals and care settings. You may have noticed these in hospitals or GP surgeries in England—but they are used globally to remind healthcare professionals about a key way to reduce hospital-transmitted infections. You can see an example of this poster in Figure 3.15.

Biosecurity in the community also forms a significant part of public health, which we will look at in greater detail in Chapter 6.

Figure 3.15 The 'Five Moments for Hand Hygiene' poster helps to remind healthcare professionals about critical times to wash their hands.

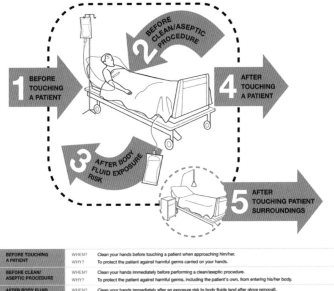

Reprinted from 5 Moments for Hand Hygiene Poster, Copyright 2009

≋ Chapter summary

- Parasites take many different forms—some multicellular and very large in size, others microscopic and difficult to detect.
- Prions are misshapen proteins which can induce similar misfolding in other proteins throughout the body. They cause progressive, degenerative, and irreversible neurological symptoms, and are very hard to detect and diagnose.

- Bacteria or fungi may be cultured in a laboratory setting to aid diagnosis.
- Parasites can sometimes be directly visualised. Blood tests for antigens against specific parasites may be useful if the parasite cannot be seen using microscopy.
- Viruses are usually impossible to culture—instead, biochemical or immunological assays, including PCR, may be used to identify different viruses.
- Biosecurity measures are designed to keep scientists and members of the public safe from potentially deadly pathogens. When these measures fail, the risk of people becoming infected rises significantly, and may even result in the release of deadly diseases into the general population.

Further reading

1. World Health Organization, *Biosafety Manual*. http://www.who.int/csr/resources/publications/biosafety/Biosafety7.pdf?ua=1

   WHO biosafety manual—gives a comprehensive review of biosafety measures which should be used by microbiology labs around the world.

2. https://mcb.berkeley.edu/labs/garcia/sites/mcb.berkeley.edu.labs.garcia/files/Teaching/2017-MCB137/Monod1949.pdf

   'The growth of bacterial cultures'—a paper from 1949, giving a good overview of the different stages of bacterial culture growth.

3. https://academic.oup.com/schizophreniabulletin/article/33/3/757/1882426/Effects-of-Toxoplasma-on-Human-Behavior

   'The effects of *Toxoplasma* on human behaviour'—an interesting look at how the amoebic parasite *Toxoplasma gondii* may affect the behaviour of infected humans.

Discussion questions

3.1 You saw in The bigger picture 3.1 that scientists are building evidence that the presence of common parasites may reduce the incidence of asthma and other allergies. If these findings are confirmed, what problems do you foresee in using this information to reduce the ever-growing rates of asthma and allergies in the UK?

3.2 Think about how you might develop a minimal medium suitable for a specific type of bacteria or fungus. How would you determine the minimal requirements for a successful culture to be grown?

3.3 Investigate bovine spongiform encephalopathy (BSE, 'mad cow disease'), scrapie in sheep, and vCJD in people. Compare the spread of the diseases and the detective work needed to identify what was happening with the story of Kuru. Why do you think BSE and vCJD have been a much bigger problem to eliminate than Kuru?

4 DEFENDING AGAINST THE INVADERS

So far, we have focused on the invaders which threaten our bodies from all sides. It would be easy to imagine that humanity has remained helpless against this tidal wave of disease, each individual taking their chances against billions of bacteria, viruses, fungi, and parasites. Fortunately for all of us, our bodies have changed and evolved alongside our enemies, and have had thousands of years to develop extremely clever defence mechanisms against all forms of invaders. Some of these defences are purely mechanical (Figure 4.1), but some are complex biochemical responses, evolved and improved over thousands of years. Some of them we share with much simpler animals, and even plants—but some of them are specific to humanity alone. Combined, these defences make up the human immune system, and they are crucial to our ability to fight off infectious diseases.

Figure 4.1 Cilia in your respiratory tract beat constantly to help remove pathogens from your lungs.

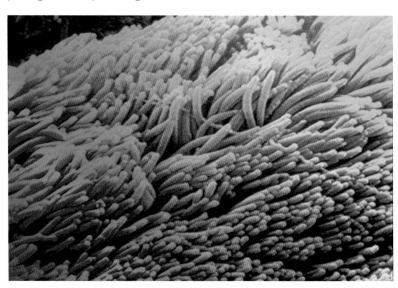

BSIP/Age Fotostock

The first line of defence—the innate immune response

Our immune systems are essentially divided in two parts. The innate immune response is the initial, basic response that the body mounts against any pathogens that it recognises. This response relies on physical barriers and simple pattern recognition, and, although it has a little flexibility to adapt, it is not as specific or as flexible as the adaptive immune response, which we will discuss later in this chapter. However, before any pathogens can trigger the internal immune response, they must first find their way into our bodies. Look at Figure 4.2; it shows the barriers that infectious agents must overcome before they can infiltrate our bodies.

The outsides of our bodies are well designed to repel the threat of many invaders. Our skin forms a tough, rubbery barrier, which is difficult for many pathogens to penetrate without the use of specialist tools. Some organisms, of course, have evolved these tools, and can burrow into our skins to infect us—but many bacteria, viruses, fungi, and parasites must look for an alternative way in. If our bodies were entirely covered in skin, it would be extremely difficult for most pathogens to infect us. Unfortunately, in order for us to breathe, to eat, to excrete waste, and to procreate, we must have passageways leading from the inside of our bodies to the open air. These orifices serve as the entry points for many of our most common infectious agents, but they are not completely defenceless. Our airways are lined with both visible and microscopic hairs, and produce mucus to trap airborne pathogens. The microscopic hairs (called cilia) then beat in one

Figure 4.2 Physical barriers help to prevent pathogens entering our bodies, and take a variety of forms.

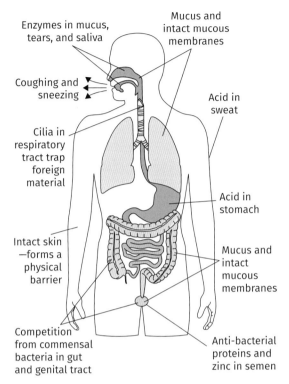

Enzymes in mucus, tears, and saliva

Mucus and intact mucous membranes

Coughing and sneezing

Acid in sweat

Cilia in respiratory tract trap foreign material

Acid in stomach

Intact skin —forms a physical barrier

Mucus and intact mucous membranes

Competition from commensal bacteria in gut and genital tract

Anti-bacterial proteins and zinc in semen

direction, moving the mucus out of our airways, where it can be coughed out or swallowed. Coughing or sneezing helps to expel pathogens from the nose or airways—which may cause other people to be infected, but helps to reduce our own pathogen load. The 'flushing' action of tears and urine can help to wash pathogens away from the eyes or bladder, removing them from the body. If a pathogen is swallowed—either enveloped in mucus from the airways, or taken in orally—it ends up in the stomach, where powerful acid and enzymes work to destroy many different kinds of pathogen. Some more specialised organisms may still end up in the gut, where more mucus is produced to attempt to prevent their infiltration into the rest of the body.

Other defences exist close to the surface, as well—the innate immune system deploys chemical defences against pathogens seeking entry to the body, and has even harnessed other microorganisms to help defend against the invaders. The skin secretes a chemical called β-defensin, which acts as an antimicrobial peptide. This peptide binds to the microbial cell membrane and forms a pore-like structure, causing essential ions and nutrients to flow out of the pathogen, and it is effective against bacteria, fungi, parasites, and even many viruses. Tears, mucus from the respiratory tract, saliva, and even sperm and vaginal secretions contain chemicals (such as β-defensins or zinc) and enzymes (such as lysozyme), all of which help to break down and destroy pathogens before they have entered the body itself.

Growing in our digestive and genitourinary tracts are large numbers of bacteria and fungi which do not cause any harm to our bodies. Called 'commensal flora', these organisms teem in their millions, and perform a number of important roles throughout our bodies—although we are only really just beginning to unravel just how crucial they may be to human health. As part of the innate immune system, commensals help to prevent infection by pathogens by competing with them for space or nutrients, and even by altering the local environment to make it more hostile for invaders. Have a look at The bigger picture 4.1—it goes into greater detail on the importance of commensal organisms in one specific area of the body.

The bigger Picture 4.1
The gut microbiome

Huge numbers of bacteria and fungi live within our bodies, teeming in their billions, mainly in the digestive and genitourinary tracts. For years, these organisms were classed as 'commensals' because their relationship with humans was thought to be a non-beneficial co-existence, but evidence is increasingly emerging that the bacteria with which we share our bodies can help fight off pathogens, and are critical to the development of a healthy immune system. Disruption to the natural balance of these organisms may even lead to the development of some autoimmune disorders. Our bodies do not exist in isolation—far from it. We are pushed and pulled, helped and hindered, by our very own microscopic world existing on and within us—our very own microbiome.

Bacteria exist in and on our bodies in almost unimaginable numbers—it is estimated that they outnumber our own cells by ten to one! About 60% of the dry weight of each human stool is made up of dead bacteria, and the living bacteria within the gut make up a significant number of the total commensal microorganisms dwelling in and on each of us. Because the gut is an environment low in oxygen, most of these bacteria are anaerobes, but some species living closer to the anus may be aerobic in nature. In total, it is estimated that between 300 and 1,000 different species of bacteria inhabit the gut, although the vast majority of those existing in large numbers come from thirty or forty distinct species. The development of this microbiome is complex, beginning during birth but continuing for the rest of our lives.

A foetus has, in theory, a completely sterile gut, with no microorganisms at all. During birth, and throughout the baby's first few days of life, it is exposed to a variety of different microorganisms from the environment around it—from its mother and family members, from medical or nursing staff who might be involved with the birth, and from other infants and people that it comes into contact with. Throughout the first few years of life, many of the organisms in the infant's gut appear to be facultative anaerobes, which help to reduce oxygen levels in the gut and pave the way for other anaerobic bacteria to thrive. This early gut microbiome varies depending on whether the infant is breast- or formula-fed, and where in the world the baby and its mother are living. As the baby's digestive tract develops over the first few years of life, it grows a layer of mucus which allows the commensal organisms to thrive. Gut-associated lymphoid tissue (GALT), crucial for the recognition of pathogens within the intestines, starts to develop at this point, and remains tolerant of the commensal organisms whilst reacting to other microorganisms. In adults, the gut microbiome is incredibly complex, and is influenced by age, diet, and geographical location. Other internal factors, such as pregnancy, may also alter the gut flora, and the use of antibiotics to treat infections will also drastically alter the numbers and types of bacteria within the gut, often allowing for overgrowth of yeasts or pathogenic bacteria due to a reduction in the number of commensals.

Our gut bacteria help to reduce the risk of infection with pathogenic agents by competing with them for space and nutrients. Their sheer number allows them to conquer many threats; their consumption of resources and their specialist nature helps them to deny a foothold to invaders. Many of the commensal bacteria also secrete antimicrobial compounds, which can slow down the reproduction of harmful bacteria. In addition to this, they appear to be able to help 'train' the immune system to release various specific chemicals called cytokines. Some species of bacteria appear to drive the production of anti-inflammatory cytokines, whereas others may stimulate production of inflammatory cytokines. These may have an effect within the gut itself, or may circulate in the bloodstream, affecting more distant areas of the body.

Although the combustion of the gut microbiome to our immune response is impressive, it is becoming clear that the role of our gut bacteria extends far beyond a merely defensive one. They help us to digest our food, breaking down compounds that our own guts simply cannot process. Without a functioning gut microbiome, our bodies would find it much more difficult to process a large number of polysaccharides—starches and sugars—as well as dietary fibre. In fact, when scientists raised rodents in a sterile environment—so they were lacking natural gut flora—they needed to eat 30% more calories to maintain their weight than control subjects with a normal gut microbiome. When all of the subjects were overfed with a high-fat, high-sugar 'Western' diet, the control group became overweight, whereas the group with a sterile gut did not. Look at Figure A below to find out how the weight of rodents with no gut flora changed once they underwent a transplant of gut bacteria from a recipient mouse who was either normal weight, overweight, or underweight.

The relationship between our guts and our physical health is clearly a very complex one. Recent research appears to show a complex relationship between our gut microbiome, the gut itself, and our central nervous system (CNS). It appears that the gut, and the bacteria within it, have the ability to regulate certain neurotransmitters and key hormones in the CNS. As we continue to unravel the mysteries of the gut and its microbiome, it is becoming increasingly clear that the microorganisms we carry within us have a profound impact on our physical and mental health throughout our lives. Discovering more about the ways our digestive tracts interact with the rest of our bodies may help us to understand diseases as diverse as multiple sclerosis, Alzheimer's disease, and depression. In the future, we may even be able to treat or prevent some diseases simply by manipulating our gut microorganmisms.

Figure A Mice raised with a sterile gut were protected against obesity when given a high-fat, high-sugar, 'Western' diet. Fascinatingly, when these mice underwent a transplant of gut bacteria from mice with a variety of different weights, they took on the characteristics of the donor, becoming overweight, underweight, or maintaining their normal weight.

(i)   **Germ-free animals are protected from high-fat, diet-induced obesity**

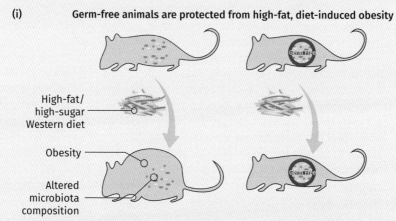

High-fat/ high-sugar Western diet

Obesity

Altered microbiota composition

(ii)   **Germ-free animals adopt phenotype of microbiota donor**

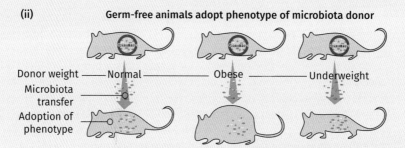

Donor weight —— Normal —————— Obese ————— Underweight

Microbiota transfer

Adoption of phenotype

From Leslie Coller et al., *Human Virology*, 5th edn (2016). By permission of Oxford University Press, Inc.

❓ Pause for thought

As we have grown to understand more about the human microbiome, it has become clear that microorganisms can have a profound effect on our health. Think about human health and disease since the introduction of antibiotics in the 1940s. Is it possible that altering our gut microbiomes through dietary changes and antibiotics could have contributed to the 'obesity epidemic' in the Western world? Do you think that 'faecal transplants' might be a useful way of helping to reduce obesity and its related diseases? If so, what resistance might this sort of treatment encounter?

The cellular response to infection

Despite our well-adapted defences, some pathogens will eventually make their way into our bodies. They may be ingested, surviving stomach acid and commensals. They may enter through a break in the skin. They could track up the genitourinary tract into the bladder, womb, or testes. Wherever they find themselves, the innate immune system still has a variety of ways in which it can combat them. Cells which become damaged or infected by pathogens stop expressing the usual proteins on their phospholipid bilayers. They release signalling molecules—cytokines and eicosanoids—which signal to the body that all is not well. Some of these chemicals—like prostaglandins, a form of eicosanoid—trigger systemic responses, like fever (a high temperature, sometimes associated with feeling hot or cold and shivering). They may also dilate blood vessels, increasing the blood supply to the infected area, and causing it to become reddened and hot to the touch. Other eicosanoids called leukotrienes attract leucocytes (white blood cells), which begin to attack any abnormal cells. Cytokines may be involved with chemical signalling, which is the case with interleukins, an important type of cytokine. There are many different types of interleukin, all fulfilling slightly different roles within the immune system. They act as signalling chemicals, affecting the migration of immune cells around the body, activating or encouraging differentiation of the various cells of the immune system. Interleukins bind with receptors on the surface of their target cells—and the effect that they generate depends greatly on the ligands involved in the binding processes, receptors expressed on the surface of the target cell, and biological cascades that are activated by the binding process. Conversely, cytokines may directly interfere with infected cells—a good example being the group of chemicals called interferons. Like interleukins, there are several different kinds of interferon, but generally they trigger nearby cells to produce an enzyme called protein kinase, which phosphorylates a protein known as eIF2, which in turn acts to slow down protein synthesis, helping to slow down viral replication.

The complement cascade

In addition to the local chemical response to invaders, the presence of foreign antigens may trigger the complement cascade, a series of chemical reactions involving over twenty different proteins within the bloodstream. Triggered by the innate immune system, complement proteins bind to the surface of a pathogen, using specific biochemical markers (an antibody attached to the cell from the immune system, the antigen expressed by the pathogen, or a chemical called mannan-binding lectin, which bonds to specific carbohydrates on the surface of certain pathogens). Once bound, a protease reaction is activated. This causes the complement proteins to cleave, releasing cytokines and triggering a further 'cascade' of cleaving reactions. These reactions ultimately have several effects: they signal to nearby white blood cells to clear any material recognised as 'foreign' by the immune system; help to increase local inflammatory effects; and activate a cell-killing membrane attack complex, which disrupts the membranes of affected cells, causing an influx of extracellular fluid and loss of crucial ions and molecules. This rapidly causes the affected cell to undergo lysis, destroying it. You can see a simplified diagram of the complement cascade in Figure 4.3.

Figure 4.3 The complement cascade, triggered by one of three possible pathways. The initial chemical reactions trigger a cascade of further reactions, resulting in the release of various proteins (C3a, C4a, C5a, C3b, C5b, C6, C7, C8, C9 in diagram) which trigger inflammatory responses or bind to their target cells, removing immune complexes or even triggering cell lysis.

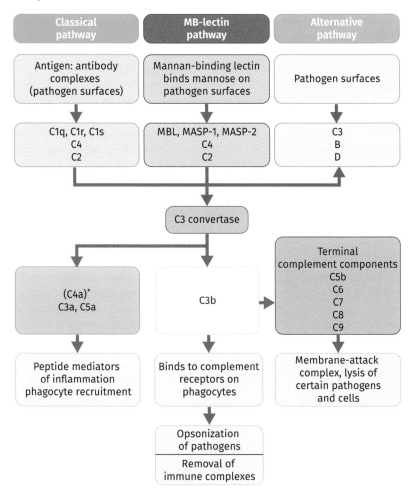

Leucocytes—white blood cell defenders

Finally, the innate immune system has a host of **leucocytes**, which all help in both identifying and eliminating pathogens. **Phagocytes** such as **neutrophils** and **macrophages** move independently through the body, searching for any antigens recognised as 'foreign'. Cytokines released from infected or damaged cells may cause these phagocytes to congregate at the site of an infection, where they engulf pathogens in a process called phagocytosis, which you can see clearly in Figure 4.4. Once the hostile cell has been engulfed, it is encased in a vesicle called a phagosome. This is then fused with a vesicle containing hydrolytic enzymes, called a lysosome. The resulting phagolysosome will destroy the pathogen, either through lysis or by releasing a burst of free radicals to damage its structure.

Figure 4.4 A neutrophil (in yellow) engulfing a bacterium (orange) through phagocytosis. The resulting phagosome will be combined with a lysosome to destroy the pathogen.

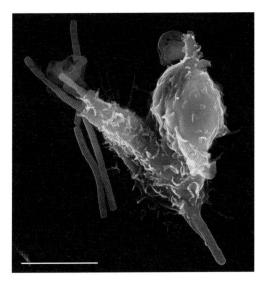

Macrophages produce a wide variety of chemicals, including cytokines, complement proteins, and enzymes, but they may also help to scavenge damaged cells or other debris, and can also act as signallers to the adaptive immune system. Mast cells—a type of specialised leucocyte—are generally found in connective tissues and mucous membranes. They help to regulate the inflammatory response, releasing histamines, lysozymes, and cytokines to trigger inflammation in local tissues. Basophils and eosinophils are other specialised leucocytes, releasing further cytokines and eicosanoids. Basophils aid in the inflammatory response, and are also involved in allergic diseases, such as asthma. Eosinophils are instrumental in fighting off parasitic infections, particularly helminths, and levels of eosinophils are commonly raised in patients suffering with parasites. Dendritic cells are specialised phagocytes which are found in tissues which are exposed to the external environment—the nose, skin, stomach, intestines, and lungs. At certain stages in their life cycle, they are shaped a little like dendritic neurons (nerve cells), with spine-like projections from their surface—but they are not related to the nervous system in any way. These dendritic cells are responsible for identifying antigens presented by pathogens, and presenting them to the adaptive immune system to allow it to mount a specific immune response. They constantly sample the surrounding environment for evidence of pathogens, using pattern recognition receptors to determine which antigens are hostile. Immature dendritic cells also phagocytose pathogens entirely, breaking their proteins down into recognisable antigens. Once they have identified a pathogen, they will then migrate to the lymph glands to present the relevant antigen to the cells of the adaptive immune system, helping to trigger the adaptive immune response.

Finally, the innate immune system possesses natural killer (NK) cells, which are also leucocytes, but are not responsible for directly attacking pathogens. Rather, these NK cells seek out and destroy damaged or compromised cells belonging to the host. They do this using a principle known as 'missing self'—looking for certain markers (such as Major Histocompatability Complex I—MHC I) present on the surface of host cells. Abnormal cells—which may be infected by a virus, or may be damaged in some other way, such as in a cancerous process—have lower levels of MHC I than normal host cells. Once they have identified a target, NK cells will then either trigger lysis of the cell using lysozymes, or mark the cell for destruction by the adaptive immune system in a process known as opsonisation. NK cells are therefore tremendously important in the immune response towards viral infections, providing a rapid immune response before the adaptive immune system can be fully activated.

Counter-attacking and remembering—the adaptive immune response

The cells of the innate immune system are hard-wired to recognise 'non-self' antigens presented by pathogens and damaged cells. They are not capable of modulating their response to different antigens, nor do they retain any 'memory' of antigens they have come across before. They are simply programmed to recognise any antigens coded as 'non-self'. The adaptive immune system, however, can recognise new antigens, and can not only respond to them at the time of infection, but can preserve a memory of them for use if the body encounters the same pathogen again in the future. This helps to provide us with a lifelong 'memory bank' of previous infections, and forms the basis of the theory of vaccination (which we will look at in Chapter 5).

The *adaptive* immune system uses elements of the *innate* immune system—both to recognise pathogens and to mount a response against them—but it also has its own specialist leukocytes, called T-lymphocytes and B-lymphocytes.

T-lymphocytes—killers and helpers

T-cells are a group of cells identified by the presence of a T-cell receptor (TCR) on the cell membrane, which is responsible for recognising fragments of antigen. This allows the T-cell to carry out its intended purpose. All T-cells are produced in the bone marrow, from haematopoetic stem cells. The immature T-cells then migrate to the thymus (a gland in the neck), where they mature into one of two main groups—helper T-cells and killer T-cells. Only about 2% of the immature T-cells end up surviving the maturation process in the thymus. The rest are destroyed as they are unable to interact properly with MHC. T-cells are also sometimes called CD4+ (for helper T-cells) or CD8+ (for killer T-cells), which refers to the type of glyco-protein expressed on the cell's surface.

Broadly speaking, helper T-cells assist killer T-cells and other parts of the immune system. They cannot destroy infected host cells or pathogens themselves, but help to 'manage' the immune response, by directing other parts of the immune system to perform these tasks. A helper T-cell in its inactivated state will wait until it is presented with a fragment of anti-gen. This might be from a dendritic cell, a macrophage, or a B-cell. Almost every cell in the body has the ability to present antigens via MHC, but these types of cell are specially designed to present antigens to the adaptive im-mune system, attaching fragments of proteins from pathogens to the major histocompatibility complex II (MHII) molecules on their cell membranes. Once they have bound an antigen to their MHC receptors, these antigen presenting cells (or APCs) will migrate to the lymphoid tissue in the body to find immature T-cells. Presenting the antigen/MHC complex to the TCR of an immature T-cell activates the cell, and it begins to divide rapidly into other helper T-cells. These helper cells all secrete cytokines, which attract other leucocytes such as macrophages and neutrophils and help to direct these cells to the areas they are most needed.

Killer T-cells are activated in the same way as helper T-cells; they lie relatively dormant until they are presented with a specific antigen, at which point they become activated and begin to rapidly divide, creating an army of cells which are designed to search for a specific antigen. Killer T-cells can also detect antigens on the surface of infected cells, via MHC I. They seek out and destroy cells demonstrating the antigen that they were activated with, releasing cytotoxic chemicals such as perforin into their target's cytoplasm to trigger apoptosis. Once an infection has been suc-cessfully overcome, the helper T-cells release regulating chemicals which help to reduce the numbers of circulating killer T-cells. There is increasing evidence to show that helper T-cells are also crucial to controlling killer T-cells, preventing them from recognising the body's own antigens as for-eign. Look at Figure 4.5—it shows the process of activating both helper and killer T-cells, and the subsequent differences between the two.

B-cells—antibody producers

B-cells are the second part of the adaptive immune response, named for the Bursa of Fabricio, where they were first discovered in birds. There are sev-eral different types of B-cells, but they all possess B-cell receptors (BCRs) on their cell membranes. Like the T-cells, they are produced in haematopoetic stem cells in the bone marrow, with immature B-cells then migrating to the spleen to mature, before lying dormant in either the spleen or the lymph nodes. Like T-cells, B-cells need to be activated before they can mount a response to infection. Once the BCR on the surface of the cell has been pre-sented with a fragment of antigen, the B-cell will absorb the antigen and

Figure 4.5 Once a T-cell has been presented with an appropriate antigen, it becomes activated—dividing rapidly. Helper T-cells will then go on to secrete cytokines and direct the immune response, whereas killer T-cells will directly attack infected cells displaying the antigen they were first presented with.

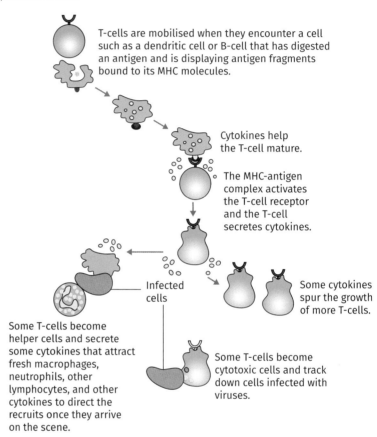

T-cells are mobilised when they encounter a cell such as a dendritic cell or B-cell that has digested an antigen and is displaying antigen fragments bound to its MHC molecules.

Cytokines help the T-cell mature.

The MHC-antigen complex activates the T-cell receptor and the T-cell secretes cytokines.

Some cytokines spur the growth of more T-cells.

Infected cells

Some T-cells become helper cells and secrete some cytokines that attract fresh macrophages, neutrophils, other lymphocytes, and other cytokines to direct the recruits once they arrive on the scene.

Some T-cells become cytotoxic cells and track down cells infected with viruses.

Wikimedia Commons/Public Domain

Figure 4.6 The basic structure of an antibody is that of a Y shape. The paratopes on the ends of the arms allow the antibody to bind to a specific antigen.

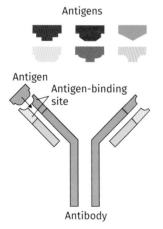

Antigens

Antigen

Antigen-binding site

Antibody

Fvasconcellos/Wikimedia Commons/Public Domain

divide it into small peptides, which are then expressed on the surface membrane of the cell, using MHC II molecules. These peptides attract a helper T-cell, which releases chemicals called **lymphokines**, activating the B-cell. The mature B-cell rapidly begins to divide, producing clones of itself called **plasma cells**. The plasma cells secrete millions of copies of an **antibody** perfectly matched to bind with the original antigen presented to the B-cell. These antibodies circulate throughout the bloodstream, binding to antigens on the surface of pathogens or infected cells. If you look at Figure 4.6, you can see the structure of an antibody. It is made up of proteins, forming two heavy chains and two light chains, which form a Y shape. At the end of each of the 'arms' of the Y there is a **paratope**—a biological lock. This is configured in such a way that it will only bind with its corresponding **epitope** on an antigen, binding them together like a biological 'lock and key'.

Once an antibody has bound to its corresponding antigen, it fulfils a number of purposes:

- Neutralisation—The antibody may block parts of the surface of the bacteria, virion, or infected cell, preventing it from carrying out its normal activities (such as cross-membrane transport). This can damage the affected cell, or prevent viruses from replicating in a host cell.

- Agglutination—Many antibodies clump together in groups, allowing them to attach to multiple antigens at once. This can 'glue' a number of foreign or infected cells together, making it easier for them to be identified and attacked by phagocytes. You can see an example of this in Figure 4.7. Antibodies may also 'coat' the outside of a pathogen, binding to multiple different antigens on the surface of the same cell. The antibodies present their 'tails' (the end of the Y) outwards, showing a region called the fragment crystallisable region (or fc region). Cells of the innate immune system (such as phagocytes, mast cells, and granulocytes) all recognise these fc regions via the use of fc receptors on the cell membrane. The recognition of antibodies' fc regions will trigger cells of the innate immune system cell to attack the antibody-coated pathogen.

Figure 4.7 A human Immunoglobulin M (IgM) molecule consists of five antibody units bound together. Each of these units has two paratopes, meaning that the entire molecule has ten paratopes, each capable of binding to a corresponding epitope.

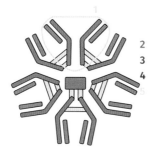

Tim Vickers/Wikimedia Commons/CC BY-SA 3.0

- Precipitation—Antibodies can attach to antigens which are soluble in human plasma (the liquid component of blood), forcing them to precipitate out of solution, and making them more attractive targets for phagocytosis.

- Complement activation—Also known as fixation. The presence of an antibody latched onto an invading cell encourages the complement cascade to attack the cell using membrane attack complex, directly causing cell lysis. The complement cascade also triggers inflammation in the local area, by attracting inflammatory cells via chemotaxis.

Long-term immunity—remembering an infection

Once an infection has been defeated by the efforts of the adaptive and innate immune systems, the body returns to watchful waiting; the specific T- and B-lymphocytes die off, and the bone marrow continues producing inactivated, 'naïve' T- and B-cells. That is not, however, the end of the story. Not all of the activated T- and B-lymphocytes die. Some of them (both T- and B-cells) differentiate into long-lived memory cells, which continue to circulate through the body, ready to begin rapidly dividing if the same pathogen should be encountered again. The B memory cells secrete antibodies, which can be measured in the bloodstream to confirm immunity to the specific pathogen. Sometimes, doctors will look for these antibodies to demonstrate that a person is immune to a certain disease—for example, to check whether a pregnant woman has previously encountered chickenpox, which can cause complications if it is contracted during pregnancy. This 'memory' of previous infections allows the immune system to rapidly produce both B- and T-lymphocytes specific to any previous pathogen that the body has encountered throughout its life, allowing a faster and more targeted response in subsequent infections. In fact, the immune system may respond so quickly to the pathogen that the infected person does not display any symptoms at all.

There are, however, only limited resources within the body, so infections which have been encountered most recently will have the highest levels of memory cells, whereas infections that the body has not been exposed to for years will have significantly lower levels of circulating memory cells. In theory, the active memory of the immune system is lifelong—although in practice, if it has been many years since the previous exposure, the host may suffer from the symptoms of infection before the immune system can increase its production of appropriate T- and B-lymphocytes. Look at Figure 4.8—the

Figure 4.8 A graph demonstrating the slow drop in antibody levels as time passes after an infection. Each peak after initial infection represents the body meeting the pathogen again. If this is a short time after the initial infection, then the increased antibody levels may not correspond to any symptoms. If it has been years since the initial exposure, the infected person may suffer from mild symptoms of the same disease, or they may remain asymptomatic (which is called inapparent reinfection).

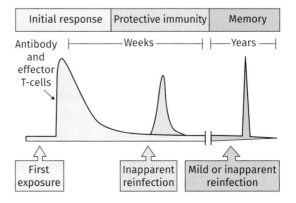

graph shows the slowly dropping levels of antibodies as time goes on, and what happens when the body meets the same pathogen again at different points along that timeline.

An immune system in overdrive—autoimmune disorders

We have already learned that both parts of the immune system use antigens and other chemicals to help them determine whether a cell they encounter is 'self' or 'non-self'. For most people, this system remains reliable through-out their lives—their immune systems remain able to accurately identify the cells of their own bodies, and only mount an immune response when they come across a cell that they identify as foreign. But sometimes this recog-nition system develops flaws. T- and B-lymphocytes can begin to target self-antigens, present on the surface of the cells within the body. When this happens, the immune system can begin to damage the organs of its own host, leading to a range of conditions which we call autoimmune disorders. The effects of the disorder will depend on which tissues within the body are attacked by the immune system—some autoimmune conditions affect only a single organ system or tissue type, such as the bowel, the joints, or the kidneys. Others may affect multiple different organ systems, and can even lead to organ failure and death, if left untreated.

Although we know that autoimmune disease occurs when the body's de-fences start attacking its own tissues, we do not completely understand what triggers this process. Sometimes, pathogens appear to express very similar antigens to some of our own cells. This can lead to the immune system correctly responding to the pathogen, but also attacking the host's cells as well. In some people, the process of 'screening' immature T- and B-lymphocytes appears to be faulty. Ordinarily, 'faulty' lymphocytes which recognise autoantigens as foreign are destroyed (a job done by the thymus, for T-lymphocytes, and by the spleen for B-lymphocytes). If this screening process stops identifying faulty lymphocytes, then these rogue T-cells and B-cells can roam through the body, attacking the host's own tissues and triggering a wider immune response through the complement cascade and cytokine response. Another theory for triggers of autoimmune diseases is called the altered glycan theory—changes in the glycans expressed on the surfaces of cells within the body which trigger an inflammatory response.

Chapter summary

- The innate immune system has a variety of defences to prevent pathogens from infiltrating the body, ranging from physical barriers to chemical defences.

- Commensal organisms living on (and in) our bodies play an important role in immune defence—but may also regulate other important aspects of our health.
- Once a pathogen has made its way into the body, it is often picked up by roaming phagocytes, which aim to destroy any tissue identified as 'non-self'.
- Macrophages such as mast cells, dendritic cells, basophils, and eosinophils aid in recognising pathogens and chemically signalling to other aspects of the immune system, such as phagocytes, T-cells, and B-cells.
- The complement cascade helps to direct the immune response, and also destroys 'non-self' cells directly via the use of membrane attack complex.
- T-cells remain inactivated until they are presented with fragments of antigens, at which point they will begin searching for cells showing the matching antigen, either attacking them directly (in the case of killer T-cells) or by using cytokines to direct other parts of the immune system to destroy the pathogen (in the case of helper T-cells).
- B-cells also remain inactivated until they are presented with an antigen. They will then produce an antibody to perfectly match this antigen, aiding the immune system in identifying and destroying pathogens.
- The T-cell and B-cell pathways are known as the adaptive immune response, as they are capable of adapting to new pathogens. They also 'remember' antigens, through the use of T and B memory cells, allowing the body to quickly respond to any pathogens that it has encountered before.
- Autoimmune disorders are caused when the immune system fails to recognise other cells within the body, attacking them in the same way that it attacks pathogens. This can give a huge variety of symptoms, depending on which type of body tissue is attacked. Autoimmune diseases may cause few symptoms, or they may make the sufferer very ill indeed.

Further reading

1. 'Gut microbiota—the neglected endocrine organism', https://academic.oup.com/mend/article/28/8/1221/2623221

   Excellent paper discussing the role of the gut microbiome in the body—discusses obesity, 'the brain–gut axis', and the potential role of the gut microbiome in mental health.

2. 'Innate immunity' from *Molecular Biology of the Cell*, 4th Edition, https://www.ncbi.nlm.nih.gov/books/NBK26846/.

   A good summary of the innate and adaptive immune systems—discusses the 'self' and 'non-self' identification process in greater depth.

3. Nan-Ping Weng, 'Aging of the Immune System—How Much Can the Adaptive Immune System Adapt?', *Immunity* (2006) 24:5, 495–9. http://www.sciencedirect.com/science/article/pii/S1074761306002214

Paper examining the decreasing immune responses associated with an aging body, particularly focusing on the decreased ability of the lymphocytes in the adaptive immune systems.

4. Ferris Jabr, 'For the Good of the Gut: Can Parasitic Worms Treat Autoimmune Diseases?', *Scientific American* (2010). https://www.scientificamerican.com/article/helminthic-therapy-mucus/

Scientific American article, which discusses the possibilities of using parasitic infections to reduce the immune response in autoimmune disorders.

Discussion questions

4.1 Look at what you have learned about the complement cascade. Why do you think that the body has evolved a biochemical cascade like this, rather than simply activating or secreting the end products of the process?

4.2 Interferons have been used to successfully treat certain infectious diseases such as hepatitis C. They have also shown promise in treating some types of autoimmune diseases, and even some cancers. Based on what you know about interferons and their role in the immune response, and further research, suggest ways in which interferons might be 'guided' to recognising and attacking specific cell types?

4.3 Why do you think our concentrations of B and T memory cells gradually decline over time? Suggest reasons why the body does not continue producing effective levels of memory cells for every infection ever encountered throughout your life.

FIGHTING AGAINST INFECTION: A CHEMICAL APPROACH

Humanity has advanced a long way in its treatment of infectious disease. For much of human history, medicine has been too primitive to enable us to cure most infections (Figure 5.1). Healers or doctors might have been able to treat some of the symptoms of the disease, keeping the patient comfortable while they suffered, but they were unable to effectively treat the root cause of their illness. Even the discovery of microorganisms and the realisation that they caused disease did not lead to any effective treatments until the twentieth century. Until this point, people suffering from any kind of infection were essentially at the mercy of their own immune systems. Either they would be able to successfully fight off the infection, or their bodies would be over-whelmed, and they would die. Even those who could successfully drive off an infection might be left permanently damaged—unable to walk or breathe due to polio, blinded by measles, or scarred by smallpox. The advances in medicine and microbiology over the past century have led to huge advances in the way we treat infectious diseases—and how we can prevent them from occurring in the first place.

Figure 5.1 In medieval times doctors had little in the way of effective treatments against infectious diseases. Leeches were thought to reduce fever by removing over-heated blood!

Wellcome Collection

Antibiotics

By the late nineteenth century, understanding of the human immune system had reached a relatively sophisticated level. However, despite understanding the causes of many different types of infectious disease, humanity still found itself with relatively few tools at its disposal to treat these infections. Antiseptic surgery focused on preventing infection, as did early work on vaccination (which will be discussed in chapter 6). A few compounds had been found to be useful for treating certain diseases—such as quinine for treating malaria—but on the whole, humanity lacked a useful toolkit to tackle common bacterial infections, from pneumonia to skin infections, scarlet fever, or meningitis.

The early part of the twentieth century saw great leaps forward, with a cure for syphilis being developed as early as 1909 in the form of a drug called salvarsan. This was tremendous news. Before salvarsan, syphilis—a sexually transmitted infection—was an incurable disease which would cause, amongst other symptoms, skin rashes, joint pains, madness, and eventually death. Without treatment, patients would suffer for years before finally dying of the disease—often also passing it on to their sexual partners. Through the 1920s and 1930s, drug companies feverishly investigated various antimicrobial compounds, discovering a class of antibiotics called the sulfonamides, which proved useful for treating various bacterial infections. Sulfonamides work by inhibiting the action of an enzyme called dihydropteroate synthetase (DHPS), which is responsible for the synthesis of folic acid inside the bacterium. Folic acid is necessary to create DNA and RNA. Inhibiting folic acid production effectively stops the bacterium from producing DNA, which stops it from dividing. Sulfonamides do not destroy the bacterium outright. They stop it from reproducing. Unlike bacteria, we human beings obtain folic acid through our diets, and do not need to synthesise it internally—so it does not have the same effect on human cells.

Sulfonamide antibiotics were used effectively to treat many different types of infection—from pneumonia to wound infections—but lack of understanding of their mechanism of action, combined with non-existent quality control and poor drug safety legislation, led to several hundred people being killed by toxic forms of the drugs. Allergies to sulfonamides were also common. In turn, this led to the creation of government departments dedicated to maintaining the safety of medicines. Sulfonamides were used extensively through the early years of the Second World War, and are credited with saving the lives of tens of thousands of people who would otherwise have died of sepsis. These early successes paved the way for one of the greatest breakthroughs in modern medicine—the discovery and development of penicillin.

Case study 5.1
Penicillin and the antibiotic era

Alexander Fleming (Figure A) is the scientist credited with the discovery of penicillin in 1928. The story is possibly one of the best-known tales of medical discovery, when Fleming serendipitously noticed that **agar plates** contaminated with a certain mould spore not only ceased to grow their

Figure A Alexander Fleming, the scientist most remembered for the discovery of penicillin.

Wellcome Collection

Figure B *Penicillium notatum*—the mould responsible for penicillin production.

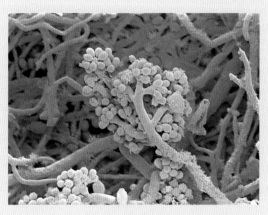

Liz Hirst, Medical Research Council/Wellcome Collection

intended bacteria, but that the bacteria were seemingly destroyed by a chemical produced by the mould (***Penicillium***; Figure B). Fleming soon demonstrated that the substance produced by the mould—dubbed penicillin—had remarkable antimicrobial properties, but his team struggled to purify the compound.

In its unrefined state, the clinical applications of penicillin were very limited, and despite attempts to use the chemical to treat superficial wound infections, Fleming's team met with little success.

Throughout the 1930s, various teams throughout the world attempted to purify penicillin into a medically useful drug, all with limited success. Either the penicillin could not be purified enough, or the resultant drug was too toxic to allow it to be used safely within the human body. Dr Cecil Paine, one of Alexander Fleming's pupils, was working at Oxford University when he tried using penicillin on a local miner. The man had a horrendous infection in his eye, caused by a stone chip, which was making him blind. Cecil washed the eye with his penicillin extract—and it got better. He also used penicillin to treat a tiny baby who had picked up an eye infection as it was born—again penicillin cured the infection and the baby's sight was saved. Paine did not write up what he had done—but he did tell a new professor at the university, a man named Howard Florey. In 1938 a team of scientists—Ernst Chain, Howard Florey, and Norman Heatley (Figure C)—found a way to purify penicillin into a form that appeared safe for use in humans. Working at Oxford University, UK, they found that even after successfully producing purified penicillin, manufacturing the drug in sufficient quantities to be medically useful still proved quite challenging. Heatley managed to make special vessels which allowed the team to grow enough mould to finally extract a useable dose of the drug. He also expressed concerns about the effects that a high dose of penicillin might have on the human body.

The first patient to be treated with intravenous penicillin was Albert Alexander, a forty-three-year-old police constable who was dying from **septicaemia** after a scratch from a rose thorn. This was exactly the opportunity the Oxford team had needed—a patient who, without treatment, would surely die. Albert was given his first dose of penicillin and started to recover well. His fevers receded, his appetite returned, and he became more lucid. But further doses of penicillin could not be made quickly enough. The Oxford university team tried harvesting penicillin from his urine, but could not obtain sufficient amounts to allow for successful treatment. Tragically, after initial improvements, Albert relapsed and ultimately died of his infection. Chastened, the scientists resolved to only experiment on sick children in future, reasoning that they would not require such high doses of the compound, and therefore might be treated for longer.

Figure C (Left to right) Ernst Chain, Howard Florey, and Norman Heatley—the leaders of the Oxford University team who managed to purify penicillin into a useful drug, ushering in a new age of treatment for common bacterial infections.

Left: Image Asset Management/World History Archive/Age Fotostock; middle: Image Asset Management/World History Archive/Age Fotostock; right: Chronicle/Alamy Stock Photo

Despite the failure in production that had led to Albert's death, the Oxford team had shown conclusively that penicillin could be a very effective way of treating even a severe infection that would otherwise have killed its host. The challenge would now be producing enough of it to use commercially—especially given that the year was 1940, and the world was in the middle of the Second World War. Drug companies weren't initially interested in penicillin. It was too difficult to synthesise, and the question of quantity remained—there was still no easy way to cheaply produce large amounts of the drug. By collaborating with American scientists, the original Oxford University team were able to improve the yields of the *Penicillium* mould. Things took off when Mary Hunt, a member of the American team, came across a mouldy cantaloupe melon on a market stall and brought it into the lab. The mould turned out to be another species of *Penicillium* mould which would grow in big fermentation tanks and gave a much higher yield of the drug. Commercial production finally became a reality. Vast quantities of penicillin were produced throughout the war years, and the drug has gone on to save millions of lives worldwide. It is still used today to treat a myriad of different bacterial infections, due to its relatively **broad spectrum** of action.

Figure 5.2 Over the past century or so, deaths from infectious diseases have plummeted. This is partly due to improved living conditions, and partly due to vaccinations (see Chapter 6)—but antibiotics have also played an enormous part.

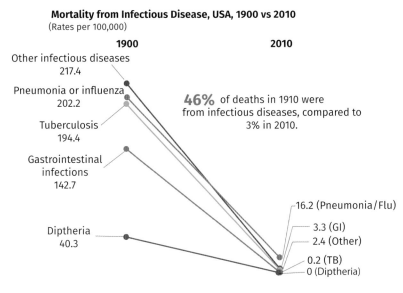

Mortality from Infectious Disease, USA, 1900 vs 2010
(Rates per 100,000)

UNC Carolina Population Center/Carolina Demography

The antibiotic age

The discovery of penicillin galvanised microbiologists, doctors, and scientists the world over. For the first time in human history, infections which would otherwise have proven fatal were treatable. At long last, humanity could strike back against the microorganisms which had caused such havoc throughout the millennia (Figure 5.2). The race was on to uncover new antibiotics, to continue marching forwards into the new antibiotic age. Streptomycin was the next antibiotic to be discovered, followed by others such as chloramphenicol, neomycin, and tetracycline.

The mechanisms by which different antibiotics have their effect are varied, but can broadly be broken down into bacteriocidal and bacteriostatic methods of action. Bacteriocidal antibiotics act to destroy the bacterium itself, causing catastrophic damage to the prokaryotic cell which cannot be repaired. Bacteriostatic antibiotics aim to slow or stop the growth of the bacteria, not killing individual cells, but preventing rapid replication of bacterial cells and allowing the immune system of the infected person to destroy the now non-replicating bacteria. Look at Table 5.1—it gives examples of many of the common antibiotics used in medicine today, along with how they work to destroy bacterial cells or inhibit bacterial division.

Choosing which antibiotic to use depends very much on what type of bacterial infection a person is suffering from, and what part of the body

Table 5.1 Some of the commonly used antibiotics in modern medicine, along with their mechanisms of action. Why do you think that some of these kill the bacteria outright, whereas others merely slow their growth?

Antibiotic class	Mechanism of action	Bacteriostatic/bacteriocidal
Penicillins (e.g. phenoxymethylpenicillin, amoxicillin, flucloxacillin)	Inhibits bacterial cell wall synthesis via competitive inhibition of transpeptidase enzyme	Bacteriocidal
Cephalosporins (e.g. cephalexin, cefutoxime, cefuroxime)	Inhibits bacterial cell wall synthesis via competitive inhibition of transpeptidase enzyme	Bacteriocidal
Vancomycin	Disrupts peptidoglycan cross-linkage	Bacteriocidal
Aminoglycosides (e.g. gentamycin, streptomycin)	Disrupts protein synthesis by irreversibly binding to 30S ribosomal subunit	Bacteriocidal
Fluroquinolones (e.g. ciprofloxacin, levofloxacin)	Inhibits DNA gyrase enzyme, disrupting bacterial DNA synthesis	Bacteriocidal
Metronidazole	Metabolic byproducts of metronidazole interrupt synthesis of bacteria DNA	Bacteriocidal
Trimethoprim/sulfonamides (e.g. trimethoprim, sulfisoxazole)	Inhibits folic acid synthesis within the bacterial cell, disrupting DNA synthesis	Bacteriostatic
Tetracyclines (e.g. tetracycline, doxycycline)	Blocks tRNA formation, preventing further protein synthesis	Bacteriostatic
Macrolides (e.g. clarithromycin, azithromycin)	Reversibly binds to the 50S ribosomal subunit, interfering with bacterial protein synthesis	Bacteriostatic
Chloramphenicol	Inhibits peptidyl transferase in bacterial ribosome, interfering with protein synthesis	Bacteriostatic
Lincosamide (e.g. clindamycin)	Inhibits peptidyl transferase, interfering with protein synthesis	Bacteriostatic

it affects. Some antibiotics—such as penicillins—will be effective against many different bacteria. These are known as broad-spectrum antibiotics, and are often used as an initial treatment, or if doctors do not know what type of bacteria they are attempting to treat. Once the type of bacterium has been isolated (usually by taking cultures of the affected area), microbiologists can get a better idea about which antibiotics are likely to be more effective. More laboratory studies can help to determine whether the bacteria have developed a resistance to any antibiotics, further guiding the choice of antibiotic used. Another important factor to take into consideration is the site of the infection, as some antibiotics are better than others at penetrating certain areas of the body. Flucloxacillin and clarithromycin are both good at penetrating skin and soft tissue, for example. Because of this, they are often used for treating cellulitis, erysipelas, and other skin infections. Metronidazole, on the other hand, is effective at combating a wide variety of anaerobic bacteria, so is often used for treating gut infections:

Antivirals

Antiviral medication has been around for a much shorter time than antibiotics, with the first antivirals only entering common usage in the 1960s and 1970s. Just as there had been challenges in developing the first antibiotics, scientists investigating early antiviral medications found that there were a number of problems involved in developing effective medication to combat viral infections.

- Viruses are very hard to study under laboratory conditions because they are obligate parasites. Unlike most bacteria, they cannot be effectively cultured, and cannot be easily observed using microscopy. This makes testing potential antiviral compounds in the laboratory very difficult.
- Viruses are often extremely good at adapting to changes in their environment. Many of them mutate rapidly, changing their protein coats, making them tricky for the immune system to identify, and making it quite challenging to develop molecules which will predictably interact with the viral envelope or capsid (you can remind yourself of viral structures by consulting Chapter 2). Antiviral medications must block viral replication completely, otherwise the virus might continue to adapt and spread through its host. This means that antiviral medications must be potent, which increases the likelihood of them being toxic to the patient.
- Many of the compounds which interfere with viral replication are also very toxic to human cells, which makes them unsuitable for use in people.
- Many viral infections are of relatively short duration—by the time the patient displays symptoms of infection, it is already too late to impact the course of their illness.

For a combination of these reasons, it has proven very expensive to develop antiviral medication compared to antibiotics, and there are still relatively few antivirals available. There are currently no broad-spectrum antiviral agents—every antiviral agent currently available has been designed to combat a single virus, or a small group of similar viruses. It is still true to say that the majority of viral diseases cannot be treated—but we have had varying levels of success at treating some of the biggest viral killers of our time, including influenza, hepatitis, and HIV/AIDS.

Antivirals, like antibiotics, work in a variety of different ways. However, all antivirals have a similar aim—to prevent replication of the virus in its host cells. Some antivirals work to block the attachment of the virion to its host, by mimicking the attachment proteins on the surface of the viral envelope or capsid. Others prevent the virus penetrating the cell membrane, or uncoating itself once it has made its way into the cytoplasm. Once the virus has uncoated itself and attempts to replicate its genetic code, we are able to substitute some of the nucleotides which would ordinarily be used by the virus to make more DNA or RNA. These substituted nucleotides work to inactivate the cell enzymes involved in DNA synthesis, preventing the virus from replicating itself. Finally, some antivirals work to prevent the virus being assembled once its constituents have been produced by a host cell. Figure 5.3 shows how various different antiviral drugs work, and which viruses they are often used to treat.

Although an individual antiviral agent may be sufficient to treat an infection, doctors will often use a 'cocktail' of antiviral medications to treat severe, chronic viral infections such as HIV or hepatitis C. This helps to ensure that, even if the virus mutates and becomes resistant to one of the drugs used, it will still hopefully be eradicated (or at least kept suppressed) by the other antivirals being taken by the patient.

Figure 5.3 Stages of viral replication targeted by antiviral medication.

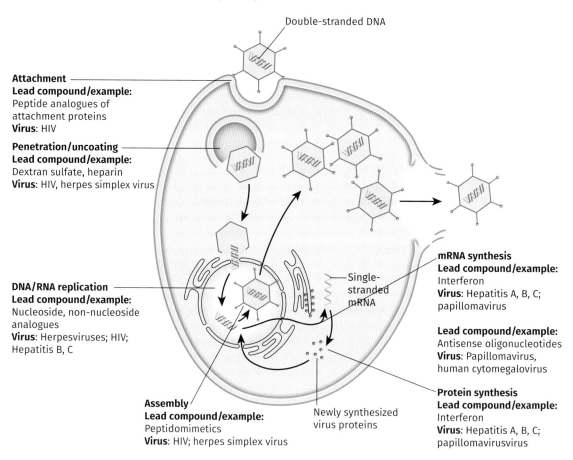

Double-stranded DNA

Attachment
Lead compound/example:
Peptide analogues of
attachment proteins
Virus: HIV

Penetration/uncoating
Lead compound/example:
Dextran sulfate, heparin
Virus: HIV, herpes simplex virus

DNA/RNA replication
Lead compound/example:
Nucleoside, non-nucleoside
analogues
Virus: Herpesviruses; HIV;
Hepatitis B, C

Assembly
Lead compound/example:
Peptidomimetics
Virus: HIV; herpes simplex virus

Single-
stranded
mRNA

Newly synthesized
virus proteins

mRNA synthesis
Lead compound/example:
Interferon
Virus: Hepatitis A, B, C;
papillomavirus

Lead compound/example:
Antisense oligonucleotides
Virus: Papillomavirus,
human cytomegalovirus

Protein synthesis
Lead compound/example:
Interferon
Virus: Hepatitis A, B, C;
papillomavirusvirus

Adapted with permission from Collier et al, Human Virology 5/e (OUP, 2016). By permission of Oxford University Press

Antifungal medications

The development of effective antifungal medication also lagged behind that of antibiotics for a number of years. The first antifungal—called nystatin—was developed in 1949. Prior to this point, the only effective treatment for fungal infections was with weak acids, and phenolic dyes. These were only effective for treating topical fungal infections, and so systemic fungal illnesses such as fungal pneumonia or fungal meningitis were impossible to treat. This was not as much of a problem as it might first appear, as systemic fungal infections are very rare, only usually affecting people with weakened immune systems.

Immediately after the Second World War, the number of people with weakened immune systems was relatively low. However, over the past sixty years, the number has risen dramatically. Advances in modern medicine have allowed people with autoimmune diseases to be treated with immuno-suppressive medication, and people who have undergone organ transplants must also remain on immunosuppressive drugs for the rest of their lives. Diseases such as HIV cause immunosuppression by attacking the immune system itself, reducing a sufferer's ability to fight infection. Chemotherapies used to treat cancer often leave people with reduced immunity as a side effect, too. Treating systemic fungal infections has therefore become much more important in the past half-century.

There are several barriers to developing effective antifungal medications. Firstly, fungal cells are eukaryotic, and bear many similarities to human cells. Therefore, many of the biochemical pathways which might be affected by an antifungal drug are also present in human cells, making the drugs potentially extremely toxic. Because systemic fungal infections are still quite rare, and new drugs are very expensive to develop, there have not been many antifungal drugs developed since the advent of nystatin in 1949—although there are now some less toxic, broader-spectrum antifungals which can be used both orally and intravenously, in addition to antifungal creams which can be used for superficial fungal skin infections.

One of the common mechanisms of action for antifungal drugs is to interfere with a group of chemicals called sterols, which form an important part of the fungal cell membrane. Animal, plant, and fungal cells all contain different sterols in their membranes, and the main sterol found in fungi is called ergosterol. Many antifungal drugs either interfere with ergosterol itself or with its synthesis. Without ergosterol, fungal cells lose their structural integrity and undergo cell death. Because human cell membranes tend to contain cholesterol instead of ergosterol, antifungals which target ergosterol are relatively safe for use in humans.

Table 5.2 shows some of the commonly used antifungal drugs, and briefly describes their mechanism of action and application. You do not need to remember the names of these drugs, but they have been included to help distinguish between the different antifungals.

Table 5.2 Common antifungals, their methods of action and uses.

Antifungal class	Mechanism of action	Route	Principal uses
Polyenes	Bind with sterols (primarily ergosterol) in the fungal cell membrane, crystallising its structure and causing leakage of cell contents	Oral, intravenous, topical (dependent on which agent is used)	Topical applications; used for oral, vaginal, or skin infections; oral or intravenous use for systemic fungal infections
Imidazoles	Disrupts the action of the enzyme lanosterol 14 α-demethylase, preventing production of ergosterol in the fungal cell membrane, so inhibiting fungal growth	Usually topical	Primarily used for skin, nail, and genital fungal infections (such as thrush)
Triazoles	Disrupts the action of the enzyme lanosterol 14 α-demethylase, preventing production of ergosterol in the fungal cell membrane, so inhibiting fungal growth	Topical or oral	Skin and genital infections; occasionally systemic infections (e.g. systemic candidiasis in immunosuppressed patients)
Thiazoles	Disrupts the action of the enzyme lanosterol 14 α-demethylase, stopping the production of ergosterol in the fungal cell membrane, so inhibiting fungal growth	Topical	Skin and genital infections
Allylamines	Inhibits the action of squalene epoxidase, another enzyme responsible for ergosterol production	Topical, oral, intravenous	Skin and nail infections, systemic fungal infections
Echinocandins	Inhibits the synthesis of glucan in the fungal cell wall via the enzyme 1,3-beta-glucan synthase	Oral (but very poorly absorbed), intravenous	Systemic fungal infections

Case study 5.2
Fungal devastation

It is easy to think of fungal infections as mild and not terribly danger-ous. Athlete's foot may be itchy and a bit unpleasant, but it's not life threatening. However, fungal infections can be deadly in the unwell, the immunocompromised, or even simply someone unlucky enough to contract a particularly deadly fungus . . .

At the 1997 Maccabiah Games in Israel, an unexpected tragedy struck the Australian team. As they entered the arena, the footbridge that they were standing on collapsed. The athletes were dumped into the Yarkon River, which had been horrendously polluted since the 1950s. Nearly seventy ath-letes were injured, and some of those pulled from the river began to show signs of serious respiratory infections rapidly after their arrival in hospital. By the following morning, seven athletes were in critical condition.

One by one the infected athletes began to die of their infection. Because no clear cause had been identified, the medical teams caring for them were powerless—antibiotics and antifungal treatments appeared to have no effect. Finally, the autopsy of one of the dead athletes revealed the pathogen—a fungus called *Pseudallescheria boydii*. This was bad news. *P. boydii* is a tough fungus, resistant to almost every known antifungal drug. At this stage, only one of the infected patients was left alive—a fifteen-year-old named Sasha Elterman (Figure A). She had been a talented tennis player, but was now con-fined to a hospital bed. The fungus had spread to her brain and spinal cord. She remained in a critical condition for months, undergoing many different operations to try to improve her condition—including multiple brain surger-ies. Sasha was still alive—but only just. She was given only a 3% chance of surviving the infection, which was overwhelming her body.

Then her medical team read about a potential new antifungal drug which had only just started clinical trials in the UK. Called voriconazole, it had only been discovered in 1991, and was a long way from getting a licence. Like the other azole antifungals, voriconazole works by disrupting the formation of fungal cell membranes, acting on lanosterol 14 α-demethylase. Lanosterol is converted by lanosterol 14 α-demethylase into an intermediate that is then converted via several other enzyme-catalysed reactions into ergosterol, which forms a critical part of the fungal cell membrane. If the fungus can-not produce this intermediate from lanosterol, it cannot go on to synthesise ergosterol—and therefore it cannot continue multiplying. In some cases it is fungistatic, and in others it appears to have a fungicidal action.

Sasha's doctors got permission to try the drug—she had nothing to lose at this stage, and without trying the new treatment, she would almost cer-tainly die. The response was dramatic. Sasha almost immediately began to improve. However, because the fungal infection had affected her so pro-foundly, she required a long course of treatment, and ultimately ended up being given the antifungal medication for 451 days. Finally, Sasha was able to

Figure A Sasha Elterman picked up a potentially fatal fungal infection when she was thrown into the polluted Yarkon River in the 1997 tragedy at the Maccabiah Games in Israel. Without a powerful antifungal drug, she might not be alive today.

Jeremy Feldman/AP/Shutterstock

leave hospital, on the road to recovery. Although she had some permanent lung damage and still suffered from seizures, she was alive. In the run up to the 2000 summer Olympic games in Sydney, Sasha was one of the carriers of the Olympic torch—thanks to a new and relatively untrialled antifungal medication. In 2002, voriconazole finally gained a licence, and is now widely used to treat life-threatening fungal infections around the world.

❓ Pause for thought

Without using voriconazole, Sasha would certainly have died. However, without knowing the potential problems involved in using a drug, is it ever completely ethical to use a drug in this way? What criteria do you feel should be applied if a new medicine is to be used before all the clinical trials are complete?

Antiparasitic medications

Antiparasitic drugs tend to be much more specific than antibiotics, antivirals, or antifungals. As you saw in Chapter 3 the term 'parasite' is very broad, covering everything from microscopic protozoa to enormous tapeworms. Because of this diversity, antiparasitic drugs tend to target a single parasite, or a group of similar parasites. Such drugs may not have any effect at all on a different parasitic organism, and so identifying exactly which parasite is the cause of an infectious disease is crucial to successful treatment. Antiparasitic drugs can be broadly broken down into three different groups—antiprotozoal drugs, antihelminthic drugs, and pesticides. Whilst the antiprotozoals and antihelminthics are usually taken internally (either orally or intravenously), pesticides are administered to the outside of the body, and are used to treat infestations with lice or scabies. Pesticides or insect repellents may also be used to protect against infection with some insect-borne diseases, such as malaria or human botfly—aiming to kill or repel the insects which carry the disease, before they can infect a human host.

Antiprotozoal agents are usually specific to one or two different species, as there are many hundreds of different types of protozoa. If you look at Table 5.3, you can see a variety of the different antiprotozoal drugs, along with their mechanisms of action, and which infections each drug is commonly used to treat.

Table 5.3 Commonly used antiprotozoal agents, their mechanisms of action, and the organisms that they are frequently used against. You do not need to remember the names of these drugs, but it is interesting to compare their mechanisms of action with the antimicrobials you have already met.

Drug	Mechanism of action	Target organisms
Eflornithine	Irreversibly binds to ornithine decarboxylase, preventing synthesis of polyamines in the target organism	Sleeping sickness caused by *Trypanosoma brucei gambiense*
Furazolidone	Believed to work by crosslinking DNA, reducing the ability of the pathogen to reproduce, and causing cell death	Used to treat *Giardia lamblia*—a common cause of protozoal gut infections
Melarsoprol	Metabolised to melarsen oxide, which binds to specific groups on pyruvate kinase, interrupting energy production in the parasite. Melarsen oxide also binds to trypanothione, inhibiting trypanothione reductase and killing the parasitic cell	Effective against sleeping sickness caused by *Trypanosoma brucei rhodesiense*
Metronidazole	Affects the synthesis of DNA in bacterial and parasitic cells	Used as an antibiotic against anaerobic bacteria, but also effective against *Trichomonas vaginalis* (the cause of Bacterial Vaginosis, or BV) and other protozoa such as *Giardia lamblia*
Paromomycin	Binds to 16S ribosomal RNA, inhibiting protein synthesis within the target cells	Amoebiasis, giardiasis, leishmaniasis, and some tapeworm infestations
Pentamidine	Not well understood—thought to interfere with DNA, RNA, and protein synthesis, depending on the organism affected	Leishmaniasis, African sleeping sickness, and *Pneumocystis jirovecii* pneumonia in immunocompromised individuals
Tinidazole	Affects DNA synthesis in bacterial and parasitic cells (similar to metronidazole)	Wide range of protozoal infections (relatively broad spectrum of action)

Figure 5.4 Most small children have threadworm infections at some stage. They are not dangerous, but they are uncomfortable and can cause sleepless nights for children and parents alike. A simple antihelminthic medicine kills the worms so they can be passed out of the body, removing the infection.

ESB Professional/Shutterstock.com

As you can see, some of the antiprotozoal drugs are also effective against some bacteria. The same is not true of the antihelminthics. Because the organisms that they target are larger and multicellular, these drugs often have completely different mechanisms of action. They can be further broken down into vermifuges (which stun their targets, allowing the body to naturally expel them) or vermicides (which kill the helminth outright, allowing the body to absorb or expel it; Figure 5.4). Table 5.4 shows the various types of antihelminthic drugs, their mechanisms of action, and their target organisms.

Compared to the large number of different drugs used to treat internal parasitic infections, there are only a few insecticides used externally to treat lice and scabies infestations. Pyrethroids (originally derived from chrysanthemum flowers, but often now synthetically manufactured) affect the nervous systems of parasites, which ingest the chemical after it is applied to the skin or hair. These drugs (such as permethrin) interfere with sodium channels to disrupt neuron function, causing spasm and death. Organophosphates (such as malathion) inhibit carboxyl ester hydrases (most notably acetylcholinesterase), again causing spasm, paralysis, and death. Ivermectin may also be used to treat lice and mite infestations, binding to glutamate-gated chloride channels, causing hyperpolarisation of the affected cells and, again, leading to paralysis and death. Of course, there are many other insecticides which might be effective against mites and lice—but these three groups have low enough levels of toxicity to allow them to be used topically on human skin, without causing systemic side effects.

Table 5.4 Common antihelminthic drugs, their mechanisms of action, and their target organisms. Once again, you do not need to remember the names of these drugs, but it is fascinating to see how they work!

Drug type	Mechanism of action	Vermifuge/ vermicide	Target organisms
Benzimidazoles	Binds to spindle microtubules, stopping nuclear division	Vermicide	Threadworms, roundworms, whipworms, tapeworms, liver flukes
Abamectin	Blocks the transmission of impulses to nerve and muscle cells by enhancing the effects of glutamate. This causes an influx of chloride ions, leading to paralysis of the organism	Vermifuge	Effective against most intestinal worm species except tapeworms
Pyrantel	Works as a depolarising neuromuscular blocking agent, causing contraction and paralysis of the affected helminths	Vermifuge	Effective against most intestinal worm species
Praziquantel	Not completely understood—thought to induce influx of calcium ions into affected cells, causing contraction and paralysis	Vermifuge	Schistosomes, tapeworms, fluke infestations
Levamisol	Nicotinic acetylcholine receptor agonist, causing continuous stimulation of parasite muscle, leading to paralysis	Vermifuge	Effective against many intestinal worm species
Salicylanilide	Inhibits glucose uptake, oxidative phosphorylation, and anaerobic metabolism	Vermicide	Tapeworm infestations

The bigger picture 5.1
Tackling malaria

In the tropical areas of the world, one of the biggest killers is still malaria. A parasitic disease caused by a number of different protozoal parasites, it infects over 200 million people worldwide, and causes nearly half a million deaths. Some charities estimate that a child is killed by malaria every two minutes. The World Health Organization (WHO) and other global health organisations are trying to tackle the problems of endemic malarial infection, but, as with any eradication programme, it is not without its challenges . . .

Figure A *Plasmodium* parasites seen under the microscope, along with human red blood cells.

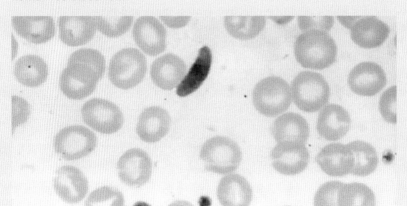

Smith Collection/Gado/Age Fotostock

Malaria is a complex disease. Caused by five different *Plasmodium* parasites (one of which can be seen in Figure A), it is transmitted to humans by mosquitoes carrying the disease, travelling through the mosquito's saliva into the bloodstream. Once the parasite—called a sporozoite at this stage—begins to circulate in its human host, it quickly finds its way to the liver, where it invades liver cells and begins to divide asexually, producing thousands of merozoites—the next stage of the *Plasmodium* life cycle. These merozoites infect red blood cells, camouflaging themselves from the host's immune system by hiding within the cells, where they too begin to multiply asexually. Once a red blood cell is full to bursting with merozoites, the parasites rupture the cell and escape to invade more red blood cells. Some of these merozoites develop into gametocytes, which continue to circulate in the host's bloodstream, waiting for another mosquito to bite the infected person. When this happens, some of the gametocytes may be taken back up into the mosquito's gut, where male and female gametocytes meet and fuse to form an ookinete. This develops into another sporozoite, which migrates to the mosquito's salivary gland, ready to infect another human host. You can see the life cycle of the *Plasmodium* parasite summarised in Figure B.

Because malaria is very good at hiding itself from the immune system (first wrapping itself in the cell membranes of liver cells, and then hiding within red blood cells), it may take some days for an infected person to start showing symptoms. The first time that the merozoites break out of the liver may be the first time that the immune system has met the pathogen, which leads to the innate immune response that you have already studied earlier in this book. The patient may suffer from fevers, malaise, jaundice, muscle or joint aches, and other flu-like symptoms. The merozoites then hide themselves within the host's red blood cells, re-emerging in characteristic 'waves' of infection, each one triggering the host's immune system again. Patients suffering from malaria will therefore often struggle with symptoms that recur every few days, mirroring the release of a new wave of parasites from their red blood cells. As the parasites damage the host's red blood cells, the patient may also suffer from anaemia.

Haemoglobin (a protein found in red blood cells) may leak into the urine from the damaged cells, staining it brown or red. As the disease progresses and the patient becomes more unwell, they may also suffer from seizures, damage to the eyes, coma, and ultimately death, as the parasite overwhelms them.

Figure B The lifecycle of the *Plasmodium* parasites responsible for causing malaria.

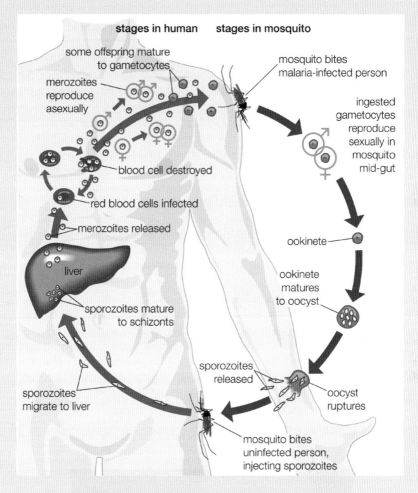

Malaria can be treated with antimalarials, but the choice of drug depends greatly on the strain of *Plasmodium* affecting the patient. Therefore, obtaining an accurate diagnosis is crucial. Ordinarily a blood sample is required to correctly identify the parasite, although sometimes the geographical area can give clues to the most likely strain involved. Once identified, the disease can often be effectively treated, and patients may make a full recovery. So, if malaria acts like a simple insect-borne protozoal infection, why is it still such a successful and deadly parasite?

The first reason is simple geography. Because malaria only affects tropical regions with large mosquito populations, it tends to occur in poorer areas of the world. This makes access to good medical care difficult for many people who are most at risk of malaria. There may be no doctor or clinic for them to access, or they may simply be unable to afford treatment. The second reason is that the disease remains endemic in the large mosquito populations in many tropical regions, passing from infected patient to mosquito and back again. It's not enough to simply treat the humans who suffer from malaria—the mosquito problem must also be tackled in order to completely eradicate the infection. The third reason is that plasmodia are extremely good at adaptation. They can change their protein coats rapidly, even during the course of a single infection, avoiding being wiped out by the immune system and making it difficult for our adaptive immunity to make effective antibodies and memory cells. The parasite hides within our own cells, shifting its protein coats to evade our defences, and often killing its hosts before they can mount an effective response against it. Of course, this ability to change rapidly gives *Plasmodium* parasites another distinct advantage against humanity—it means that malaria is very difficult to vaccinate against, and that the parasites are equally good at developing resistance to any antimalarials used against them. Drug-resistant malaria is a significant problem in all areas where malaria is endemic, and there are now some strains in Southeast Asia which appear to be completely resistant to any existing antimalarials.

The fight against malaria cannot be won simply by treating the infection once it has taken hold in an individual patient. The rates of infection are too high, and the parasite is too successful, for this to work. Instead, health authorities must take a combined approach to eradicating the disease, looking at what can be done to prevent infection in the first place, and conserving antimalarial treatments for those who have confirmed infections. Some of those measures are simple, but surprisingly effective. Mosquito netting (Figure C), for example, can drastically reduce the rate of malarial infection, especially if the nets are impregnated with insecticides. They are also a very cheap measure to implement, meaning that governments and charities can obtain them easily and distribute large numbers of them. A single mosquito net can protect multiple people from bites, and is re-usable. This makes them a very cost-effective measure for tackling the disease.

Figure C Simple and relatively cheap measures like insecticide-impregnated bed nets can have a big impact on the incidence of malaria.

Olivier Asselin/Alamy/Age Fotostock

Use of insecticides, particularly in areas with lots of standing water, can also be very effective at reducing the local population of mosquitoes. The WHO estimates that if 80% of the homes in a malaria zone are sprayed internally with insecticides, rates of malaria can be significantly reduced. However, care must be taken with insecticides, as many of them are also highly toxic to other organisms, and can cause catastrophic environmental damage if they enter the food chain. This means that widespread spraying of the local environment to eradicate mosquitoes is not encouraged. In addition to this, many species of mosquito are beginning to develop a level of resistance to insecticides, making effective use more challenging. Because mosquito larvae can only survive in water, some countries attempt to prevent areas of stagnant water from building up. This can involve draining ponds and ditches, and even fining individual people for leaving buckets or basins full of water standing around. All of these measures can help to reduce the number of mosquitoes, and their ability to spread disease. The WHO suggests that this should be the mainstay of attempting to tackle malaria worldwide.

Although vaccinating against malaria is difficult because of the different strains of *Plasmodium* and their rapidly changing nature, it is not impossible. Since 2015, limited clinical trials of a vaccine against malaria have been carried out. Mosquirix comprises a portion of sporozoite protein which has been fused with a hepatitis B surface antigen (RTS) and combined with another hepatitis B surface antigen (S). Together they make up a structure called a virus-like particle, designed to trigger the adaptive immune system into antigen recognition, antibody manufacture, and memory cell activation. In addition to this, the vaccine contains an **adjuvant** designed to enhance adaptive immune response and stabilise the antigens. In this case, the adjuvant comprises an extract from the bark of *Quillaja saponaria*—also known as the soap bark tree—and monophosphoryl lipid A. The vaccine appears to give partial immunity to malaria for young children, and is still undergoing clinical trials to further determine its efficacy and safety. The WHO states that it may become an important part of malarial control measures, but that at present it does not look as though it will replace other preventative and curative treatments. The fight against this disease goes on, but it does not look as though we will be able to eradicate it for some time yet.

Chapter summary

- Antibiotics are either bacteriostatic (halting division and growth of bacteria without killing them) or bacteriocidal (acting to destroy bacteria directly). They have a number of different common mechanisms of action, with many antibiotics working against multiple different types of bacteria.

- Antiviral medications must work to inhibit all reproduction of the virus, otherwise resistant mutants will quickly appear.
- Many viral infections cannot be treated with antiviral agents, as each drug must be very specific to the individual virus (or small group of viruses).
- By the time symptoms have occurred, it is often impossible to affect the outcome of a viral infection by using antiviral medication, and therefore most common viral infections are not treated using antivirals.
- Antifungals are challenging to develop, as fungal cells share many similarities with the cells in the human body. Antifungal agents work either by inhibiting fungal growth or by destroying fungal cells directly.
- Antiparasitic agents are usually quite specifically tailored to individual parasites, or groups of similar parasites. Identifying the organism causing a parasitic infection is therefore crucial to successful treatment.

Further reading

1. 'How Antibiotics Kill Bacteria: From Targets to Networks'. https://www.ncbi.nlm.nih.gov/pmc/articles/PMC2896384/

   A paper giving greater detail on the mechanism of action of a variety of different classes of antibiotic.

2. ABPI, 'Making Medecines'. https://www.abpischools.org.uk/topic/makingmedicines/1/1

   Making drugs to combat infectious diseases—or indeed making any drug—is a long, expensive process—find out more about how we develop new antimicrobials here.

3. World Health Organization, 'Global Infectious Disease Surveillance'. http://www.who.int/mediacentre/factsheets/fs200/en/

   Information from the WHO on their role in disease surveillance, and their sources of information.

Discussion questions

5.1 Antibiotics, antivirals, antifungal medications, and antiparasitic drugs must go through long periods of testing before they are deemed safe to use in humans. Suggest ways in which their relative safety can be determined before they are given to human subjects.

5.2 A new tool available to scientists is the ability to analyse the genome of pathogens quickly and easily. Looking at the mechanisms of action of many of the drugs mentioned in this chapter, discuss the impact this new technology might have of our ability to treat infectious diseases.

6 FIGHTING AGAINST INFECTION: VACCINATION AND PUBLIC HEALTH

Although antimicrobials and antiparasitic drugs can be very effective at treating a disease once it has taken hold within its host, our own immune systems are often more than capable of handling infections on their own. As you have seen in Chapter 4, once we have been exposed to a pathogen, our body retains a memory of the antigens presented by that particular pathogen, circulating the body in the form of B and T memory cells. If we are exposed to that same pathogen again, the body is usually able to quickly produce B- and T-lymphocytes specific to the pathogen, often fighting off the infection before it produces anything more than the mildest of symptoms. This, of course, relies on surviving the initial infection with the disease. But what if we could introduce our bodies to the concept of the disease, a weakened form of infection that might stimulate an immune response without triggering the symptoms of a full-blown infection? That is the basic principle behind **vaccination**. And vaccination is part of a bigger picture—it may be one of the most successful aspects of **public health** programmes worldwide (Figure 6.1).

Figure 6.1 Around the world, vaccination is being used to reduce and even eliminate infectious diseases which have killed thousands and even millions of people in recent history.

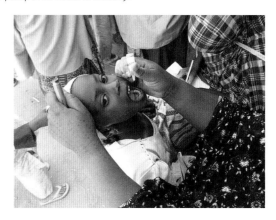

Photo: Alford 'A.J.' Williams. Public Health Image Library (PHIL)/CDC/Molly Kurnit, M.P.H.

Vaccination

Look at Figure 6.2: this individual is infected with smallpox, a viral illness which causes hundreds of painful blisters to form, as well as making the sufferer very ill indeed due to the body's own inflammatory responses. You may have never heard of smallpox—and with good reason, because it no longer exists. Thanks to a dedicated vaccination programme during the nineteenth and twentieth centuries, the disease has been wiped out (Figure 6.3). A few samples still exist in laboratories, for ongoing study, but to all intents and purposes smallpox is now extinct 'in the wild'. The last known natural case occurred in 1977, in Somalia, and in 1980 the World Health Organization (WHO) declared the disease to have been eradicated. Smallpox appeared around 10,000 years BCE, scarred Egyptian pharaohs, quite probably contributed to the decline of the Roman Empire, and played a part in the fall of the Aztecs and the Incas in the Americas. It killed millions of people throughout history and caused disfiguring scarring to many of its survivors, but it has been successfully extinguished by vaccination.

Figure 6.2 This person is suffering from smallpox, a deadly and disfiguring disease which has now been successfully wiped out by vaccination.

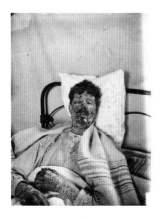

Wellcome Collection

How does vaccination work?

The principle of vaccination is a relatively simple one (Figure 6.4). By exposing the body to a weakened strain of disease, the adaptive immune system is able to produce antibodies (and memory cells) to the specific antigens presented by the infection. The exposure might be to a living (but weakened) strain of disease, a dead (but intact) version of the pathogen, modified toxins, or even antigen-bearing fragments of the pathogen—different vaccines use different approaches. Because the infectious agent is weakened, the person being vaccinated does not suffer from the full effects of

Figure 6.3 The impact of vaccination on deaths from smallpox in Sweden between 1722 and 1843.

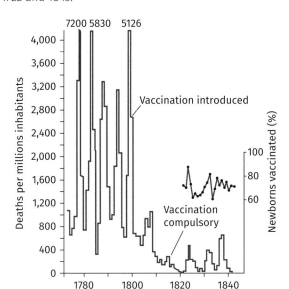

the illness, but is able to produce antibodies to the threat. The person who is vaccinated may suffer from some mild symptoms of the infectious disease, but is then protected from contracting the infection in the future because of the memory cells of the adaptive immune system. If the body ever encounters the same pathogen, the memory cells can swing rapidly into action, and an army of T- and B-lymphocytes already coded to attack the pathogen, identified by its unique antigens, can quickly counter the infection before it can cause any symptoms.

Figure 6.4 Vaccinations harness the power of the immune system to protect us against potentially lethal diseases.

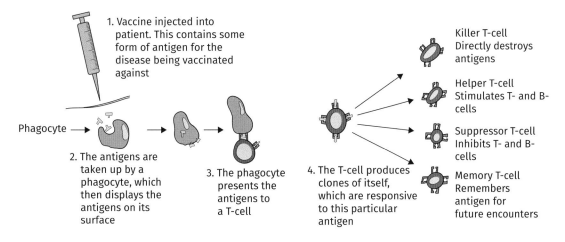

Scientific approach 6.1
The history of vaccination

There is good evidence that the ancient Chinese used vaccination against smallpox—inoculating individuals with pus isolated from smallpox sufferers. However, written reports from the time are unreliable. It is likely that other cultures around the world developed similar methods of inoculation based on the knowledge that an individual who had survived a smallpox infection was then immune from catching the disease again. This crude form of immunisation appears to have arrived in Europe during the early 1700s—possibly brought by travellers from modern-day Istanbul. By 1714, the Royal Society in London had been made aware of the technique by European scientists, but these revelations did not change the minds of the medical establishment in England, and 'variolation' did not become widely used in England until the intervention of a remarkable woman—Lady Mary Wortley Montagu (Figure A).

In 1715, Lady Montagu had suffered from a bout of smallpox, which left her severely disfigured; the disease also killed her brother. Not long after this, her husband was appointed as the ambassador to the Ottoman Empire, so they had to travel to Istanbul. Soon after they arrived, Lady Montagu wrote a letter to a friend of hers, discussing the technique of variolation, and its relatively widespread use amongst the local people. She went on to become so convinced by the effectiveness of variolation that she ordered the embassy surgeon to perform the procedure on her five-year-old son. He survived, and

Figure A Lady Mary Wortley Montagu, who brought variolation to the UK, paving the way for the work of Edward Jenner.

Wellcome Collection

on their return to England in 1721, she ordered the same surgeon to inoculate her four-year-old daughter, in front of doctors from the Royal Court.

Word started to spread about this new procedure, and before long the embassy surgeon—a man named Charles Maitland—was given a royal licence to perform a trial of variolation on six prisoners in Newgate Prison. The experiment went ahead in August 1721, observed by court physicians, members of the Royal Society, and doctors from the Royal College of Physicians. All of the prisoners survived the experiment, and those who were later exposed to smallpox were proven to be immune. Maitland repeated his experiment on a group of orphans, before performing variolation on the children of the Princess of Wales. After this last success, the procedure gained more widespread acceptance.

Variolation spread rapidly throughout Europe, gaining popularity as a method of protection against smallpox. The procedure wasn't perfect—it still carried the risk of transmitting smallpox, which could kill the individual or lead to an outbreak of disease. Because the pus used for variolation was taken directly from the sores of an infected person, it was also possible to transmit other diseases along with the smallpox—including deadly infections such as syphilis or tuberculosis. However, variolation was still significantly safer than smallpox, carrying only a 2–3% mortality rate versus the 14% (or even higher) mortality rate of the naturally occurring disease. Many people underwent the procedure, including many members of the European royal families. Some rulers even inoculated their armies, on the understanding that a healthier fighting force was a more effective one. The practice of variolation also spread to the New World, and two American doctors were able to prove its effectiveness during a smallpox outbreak in Boston during 1721. Mather and Boylston were able to demonstrate a reduction in mortality from 14% to just 2% in those who had been variolated—but some of the general public were not convinced. At the height of the controversy, Mather even had a bomb thrown through the window of his house! Despite misgivings from some scientists and members of the public, variolation continued to be performed throughout Europe and the USA. In 1757, a group of children in Gloucester underwent variolation. Amongst them was an eight-year-old boy named Edward Jenner, who would be instrumental in taking the next step in immunisation.

Jenner went on to become a doctor—and the man who is credited with the scientific development of vaccination, rather than variolation (Figure B). As a keen natural scientist as well as a doctor, Jenner was the first person to conduct scientific trials into the effectiveness of inoculation, publishing a paper in 1796 proving the efficacy of his technique, which he dubbed vaccination. He had noticed that milk-maids never appeared to catch smallpox, but would often have suffered from cowpox—a much milder infection which causes blisters and sores, but does not have the same severe consequences as smallpox. Jenner surmised that these two facts might be related. He was so sure that his theory was correct that in May 1796 he inoculated a healthy eight-year-old boy, James Phipps, with pus drawn from a cowpox blister from Sarah Nelms, a young dairymaid. The boy had a mild illness for a few days, but quickly recovered. In July 1796, Jenner inoculated the boy again—this time with pus from a fresh smallpox lesion! The boy remained completely healthy, allowing Jenner to prove that the child was immune to smallpox thanks to the cowpox inoculation. Told that he needed more data to publish his theory, Jenner went on

Figure B Edward Jenner, the English scientist credited with the invention of the term 'vaccination', and pioneer of smallpox vaccination in Europe.

Wellcome Collection

Figure C It wasn't all plain sailing. Many people—especially religious groups—did not accept Jenner's work on vaccination. He had to deal with abuse and ridicule (demonstrated in this cartoon) as well as praise and honours.

Wellcome Collection

to perform the experiment on a number of other children—including his own eleven-month-old son—with the same result. Eventually Jenner published a booklet himself on vaccination, and the practice spread relatively quickly (despite some opposition; see Figure C)—because it worked and (unlike the older technique of variolation) did not risk giving people smallpox.

In the early 1800s, the British parliament awarded him £30,000, a huge sum at the time, in honour of his work. Jenner always remained a practising doctor. He even built a hut in his garden where he vaccinated poor people for nothing—at a time when everyone had to pay for health care, or go without it.

❓ Pause for thought

The work of both Lady Mary Wortley Montagu and Dr Edward Jenner would be not only impossible today, but also viewed as highly unethical. Yet between them they have saved countless lives. Discuss the differences in attitudes to medical experimentation and development in the eighteenth century and today.

Herd immunity

In the latter half of the nineteenth century, scientists began to develop vaccinations against other common infectious diseases—rabies, anthrax, plague, typhus, and cholera. Louis Pasteur, the famous French scientist you met in Chapter 1, was one of the prime movers in this, with his classic work on vaccines against anthrax and rabies. Despite not having access to antimicrobial agents, humanity finally had a defence against some of the most virulent killers in history. However, full understanding of the science behind vaccination lagged behind these discoveries, as did the ability to successfully defend against infectious disease on a societal level. A vaccine might protect an individual or family against infection, but there were no national or international vaccination campaigns to eradicate disease locally or globally.

A key concept when discussing vaccination is that of herd immunity. Very few vaccines will be 100% effective—some people will not become immune to the disease, despite being vaccinated. It is estimated, for example, that the vaccine for tuberculosis (the BCG) has between a 60% and 80% efficacy at preventing serious tuberculosis infections, and a variable rate at preventing pulmonary tuberculosis depending on geographical area and age at vaccination. However, even if a vaccine is not completely effective at preventing the disease in an individual, it can help to prevent the spread of disease from person to person.

Look at Figure 6.5—it demonstrates how the concept of herd immunity helps to prevent infectious disease at a societal level. Essentially, even a vaccine that is not 100% effective at preventing disease in an individual

Figure 6.5 Herd immunity helps to contain the spread of infectious diseases, reducing the risk of individuals spreading disease through a population. The top box shows a community in which few people are immunised. The disease can spread effectively from person to person. The middle box shows a community in which a few people are vaccinated. These vaccinated people may be protected from the infection, but the disease is still free to spread easily through the population of non-vaccinated individuals. The bottom box shows a community in which most of the population have been effectively vaccinated. The disease cannot easily spread, even to those few individuals who are not vaccinated, because the vaccinated individuals are able to mount an immune response to the disease without becoming symptomatic or passing the disease on to those around them.

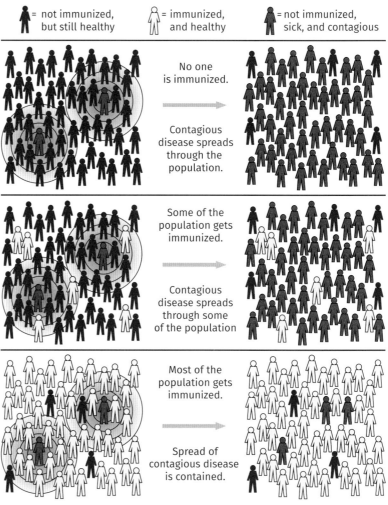

T Karcher/Wikimedia Commons/CC-BY-SA-4.0

can still be tremendously effective at reducing the rates of transmission through a group of people. This can help to minimise exposure to a disease for people who are particularly vulnerable, such as the very young, the very elderly, or the immunosuppressed. The proportion of the population who need to be immune to an infectious disease to give effective herd immunity depends on the pathogen involved. For example, measles is highly infectious, and easily passes from person to person. One individual infected with measles can readily pass the infection to ten or fifteen others. Because measles is so infectious, herd immunity is only achieved if 90–95% of people in a population are vaccinated. Other diseases—such as polio—are harder to spread from one individual to another. This means a lower proportion of the population need to be vaccinated before herd immunity is achieved—for polio, only 80–85% of the population needs to be vaccinated to give herd immunity against the disease.

Throughout the twentieth century, vaccinations for many common infections were developed. Increasingly effective public health programmes throughout the developed world helped to ensure that, at the beginning of the twenty-first century, we suffer from far fewer of the infections that once plagued our ancestors. Diseases such as measles, whooping cough, polio, or diphtheria—which were often fatal in childhood—are now thankfully very rare in the developed world, where we have advanced programmes of vaccination throughout our childhoods. Thanks to the efforts of international health organisations like the WHO and the work of many thousands of workers giving polio vaccinations across the globe, polio has been almost completely eradicated. In 2016 there were only thirty-six reported cases of polio worldwide—a significant drop from the estimated 350,000 cases in 1988.

The case for vaccination seems fairly clear-cut. It offers a way to drastically reduce infection rates society-wide, usually with a very low complication rate. Most vaccines are incapable of causing the diseases that they are vaccinating against, and the worst side effect that people suffer from is a slightly sore arm from the injection. Occasionally some vaccines may provoke a mild immune response, leading to feverish symptoms and feeling mildly unwell. Very rarely, people may suffer from an allergic reaction to one of the constituents in the vaccine. However, vaccines are generally very safe. There are, however, a small but vocal minority of people opposed to vaccination, for a number of reasons. Have a look at The bigger picture 6.1—it shows how easily public perception of an issue can be changed by poor scientific, and unscrupulous scientists, and poor journalism.

The bigger picture 6.1
Andrew Wakefield—a doctor disgraced

In 1998, a British doctor named Andrew Wakefield (Figure A) published a paper that made a staggering suggestion. Wakefield claimed that he had uncovered a link between the MMR vaccine (routinely used in the UK to vaccinate against measles, mumps, and rubella) and autism, a lifelong developmental disorder which affects how sufferers view the world and interact with others. Wakefield claimed that children vaccinated with the MMR were

Figure A Andrew Wakefield, responsible for damaging the public perception of vaccinations through fraudulent research. Children became ill and some even died as a result of his claims.

Photo by Daniel Berehulak/Getty Images

more likely to suffer from autism and bowel disorders, and that the vaccination was the likely cause for this. The response was immediate, and alarming. Newspapers reported Wakefield's study as being the truth, and the public perception on the MMR—and vaccinations in general—started to shift.

Unfortunately, Wakefield's study was not only wrong, it was fabricated. He had only studied twelve children with autism, and much of the paper relied on parental recall and beliefs. He had taken blood samples from children at a party without parental consent, paying the children a small amount of money to do so. Two of the children in his study had not even had the MMR vaccine. In fact, the paper itself did not claim a causal link between autism and the MMR—but at a press conference before the paper had been published, Wakefield sensationalised the research, claiming a clear link between the two. It has been suggested that this approach of 'science by press conference', and the exaggeration of the significance of his research, is what caused the international panic about vaccinations.

Immediately after the publication of Wakefield's paper, other researchers around the world began setting out to see if his results could be replicated. They could not—no causal link between the MMR vaccine and autism could be found. But the damage had been done, and vaccination rates started to drop in the UK and worldwide. In 2003/2004, MMR vaccination rates in the UK were as low as 80%—significantly lower than the 95% level recommended by the WHO for effective herd immunity. The government, public health officials, and medical professionals all attempted to reassure the public that the link made by Wakefield did not exist, and that there was no link between autism and the MMR—but their responses have been criticised for being ineffective.

Eventually, Wakefield was brought to justice. His research was found to be fraudulent, and he was also found to have applied for a patent for a single-dose measles vaccine, which raised significant questions about his motives for discrediting the MMR. He had ignored data which had not agreed with his

Figure B Many people do not understand how vaccinations work—and parents are always protective of their children. The MMR scare was not the first and may not be the last. This graph shows the impact of a scare about the whooping cough vaccine in the twentieth century—and the impact it had on the incidence of the potentially deadly disease.

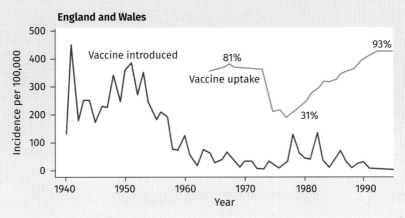

Contains public-sector information licensed under the Open Government Licence v3.0

theories, had fabricated results, and left out important facts about the children he had studied. His research, therefore, was entirely flawed, motivated by the possibility for financial and personal gain. It also emerged that he had been paid by teams of lawyers to undermine the MMR vaccine. Wakefield was stripped of his licence to practice medicine in the UK, and moved to the United States. To this day he appears to maintain his belief that vaccines are responsible for autism, and that he has been unfairly framed for 'any attempt to investigate valid vaccine safety concerns'—although analysis of data from a large number of studies into vaccine safety, involving many thousands of children in countries all over the world, have demonstrated none of the effects reported by Wakefield in his paper or any of his other published literature.

❓ Pause for thought

- *Wakefield's research was not only fraudulent, it appears to have been motivated by personal greed. Whilst researchers will often have a passionate interest in their topics, it is obviously tremendously important that the results of their research are completely unbiased, and that the scientific methods used are ethical, transparent, and accessible to others who might want to read the resulting paper. If a hypothesis is useful, then it should also be possible for others to replicate the study and get similar results. How do you think we can we ensure the integrity of scientific research, and reduce bias introduced by promises of personal or financial gain?*

- *Investigate the UK whooping cough scare of the 1970s—why did it happen, what was the effect on public health, and what is the uptake of the whooping cough vaccine now? Look at Figure B. How would you use this data to argue for the efficacy of vaccination against infectious disease?*
- *Think about the power that social media now holds over people's lives. If a modern-day Andrew Wakefield were to make similar claims over another vaccine, do you think that the spread of information (and misinformation) would show a different pattern? Do you think it would be easier or more difficult to undo the damage done by such a person?*

Overcoming animosity

Wakefield was not the first person to make claims that vaccines were harmful. Anti-vaccination movements have existed since the early 1800s, when smallpox vaccination programmes in the UK led to protests that vaccination was 'unchristian', rooted in deep religious beliefs and suspicion of the medical establishment. Many people at the time did not understand or believe the germ theory of disease, and vaccination seemed barbaric to them. A simmering undercurrent of anti-vaccination sentiment has persisted throughout the twentieth century and to the present day, with claims like Wakefield's adding fuel to the fire of public mistrust. Concerns over preservatives such as thiomersal (a compound containing mercury) in vaccines and links to autism—again, unproven by any research—led to thiomersal being removed from vaccines in the UK and USA.

Concerns for the health of their children and mistrust of the medical establishment, or the science behind vaccination, has made some parents more reluctant to have their children vaccinated. Claims that are often made by anti-vaccination proponents are that doctors are paid by pharmaceutical companies to vaccinate needlessly (which is *not* true), that improvements in public health and sanitation have reduced rates of infectious disease, not vaccinations (they have played an enormous part—see below—but vaccinations are also an important part of those public health changes), or that vaccines are filled with 'toxins' which cause harm (which is untrue and meaningless—the word 'toxin' simply meaning a poison of plant or animal origin. Some vaccines do indeed contain modified versions of bacterial toxins—but these have been altered so that they still trigger an immune response without harming the person who is given the vaccine. Some people argue that they suffered from illnesses like measles, and that they survived with no ill effects—forgetting that only the survivors of these illnesses are now here to tell their stories.

In a way, vaccines are the victim of their own success. In most developed countries of the world, people no longer expect to lose at least one child to infectious diseases such as diphtheria, whooping cough, or measles. So our awareness of the horrors of the diseases that vaccines protect us from has faded, leaving space for suspicion and concerns about the vaccinations themselves. In some areas of the world, people still expect to lose children to preventable infectious diseases. For example, a common saying in the developing world is 'Don't count your children until they have had measles.' In other words, you do not know how many children you will be left with until after they have suffered from and survived this common killer. Here in the UK, deaths from measles are now almost unheard of, thanks to the measles element of the MMR vaccine.

Often people opposed to vaccination are difficult to reach through public health awareness campaigns, or are very resistant to changing their views. This presents healthcare professionals with a difficult choice, as it is often the children of these people who end up unvaccinated. Of course, auton-omy—the ability to choose—is a tremendously important part of medicine. But a child is too young to be able to choose whether it is vaccinated or not, and vaccination programmes in the UK take place in the first few years of life. For a vaccination programme to be truly effective, enough people need to be immunised to confer herd immunity to the population. Governments and public health officials must make a difficult choice—do we respect the wishes of the individual who does not wish to vaccinate their children, or do we sacrifice some freedom of choice for a 'greater good'? Currently it is not mandatory to vaccinate your children in the UK. In other parts of the world, governments have started to penalise parents who choose not to vaccinate—for example, in some parts of the United States and Italy chil-dren cannot attend school unless they have been vaccinated.

Public health—preventing, tracking, and treating

The discipline we now call public health has existed for millennia. Looking back at ancient Chinese medicine, to the ancient Greeks and Romans, the ancient Arabic world, and even to medieval theories about health and disease, it is very clear that for many thousands of years people have understood that disease could spread from person to person. Theories may have varied as to how or why this spread occurred, but many ancient civil-isations attempted to enact some measures to prevent the unchecked spread of infection. From providing spittoons and rubbish collectors in ancient Chinese cities, to public baths and sewage systems (Figure 6.6), to the isola-tion and quarantine of houses, districts, and even entire villages in plague-ravaged Europe, doctors, officials, and the general public have long tried to keep themselves safe from the threat of infectious disease.

By the 1700s—after the waves of plague decimating Europe had begun to subside somewhat—officials and physicians had started to take an interest in charting the populations of their countries, using census data to look at birth rates, death rates, diseases, and demographics. This marked the earliest evidence of the science of epidemiology, at least in the Western world. However, before the industrial revolution and subsequent mass urbanisation of much of Europe, governments did not see a need to involve themselves in health care beyond a very basic level. Most hospitals were run as charities or businesses, and any public healthcare measures introduced were purely to prevent diseases as they occurred, rather than aiming to keep citizens healthy and disease-free.

Figure 6.6 Throughout history, humans have tried to keep themselves safe from the threat of infectious disease. This spitoon and photograph of ancient sewer tunnels are good examples of early public health measures.

Left: © Walters Art Museum/Wikimedia Commons/CC BY-SA 3.0; right: Glyn Thomas Photography/Alamy Stock Photo

The cholera effect

England experienced a dramatic change in the nineteenth century, when cholera found its way into the country. A disease caused by the bacterium *Vibrio cholerae*, cholera produces violent stomach cramps, watery diarrhoea, and vomiting in its victims. Without prompt treatment, patients quickly become dehydrated and die—their eyes sunken, their skin pallid and cold, and their hands and feet furrowed and wrinkled from the rapid loss of fluids and electrolytes. Young children and the elderly are often the worst affected, and it is easy to transmit the infection from person to person, with millions of cholera bacteria being present in the vomit and diarrhoea of the infected. Without good hand hygiene and access to both clean water and decent sewerage systems cholera spreads rapidly, causing symptoms in a new host within a few hours of being introduced. In the crowded slums of Britain's cities, the disease prospered and spread with little to hold it in check. This attracted the attention of various prominent public health activists at the time, and forced the government to start taking steps to ensure the health and wellbeing of its citizens, particularly the poorer and less educated ones.

Programmes were introduced to supply clean drinking water and adequate sewerage to the cities, but progress was slow, and the end of the 1800s still saw many miles of inadequate slums in cities throughout the UK, with little access to decent nutrition or clean water. Indeed, it was not until after the Second World War—when many areas of the larger cities had been destroyed by bombing raids—that many of these cramped, inadequate houses were eventually destroyed, and their residents re-housed in newer, arguably more suitable accommodation. Other public health programmes, however, proved more effective and swifter-acting—such as the development of vaccination against disease.

An English doctor named John Snow is now considered to be one of the fathers of modern epidemiology. In 1854, an area of London around Broad Street was in the grip of a cholera outbreak, that killed over 150 people in the first few days. Two weeks later, nearly 500 people had died—one of the worst cholera outbreaks in London's history. The miasma theory of disease still held sway in academic circles, and Pasteur would not publicly reveal his germ theory of disease until the 1860s. John Snow and a local reverend named Henry Whitehead were suspicious that the cause of cholera was not, in fact, due to foul air, and that the disease was spread by drinking contaminated water. By interviewing local residents, Snow and Whitehead were able to prove that a water pump on Broad Street was the likely cause of the outbreak. Snow would later create a dot map, clearly showing that cases of cholera were clustered around the Broad Street water pump. You can see the map in Figure 6.7.

Figure 6.7 Snow's dot map shows cases of cholera clustered around the Broad Street water pump, London. Black squares marked on the map show cases of the disease.

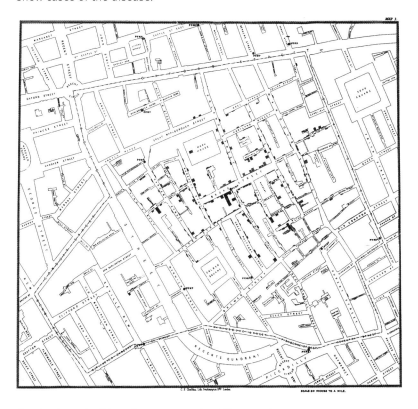

Although the organism causing the cholera outbreak was not conclusively identified, Snow had good evidence that the Broad Street pump was the source of the disease. His statistical analysis was robust enough for the local parish authorities to disable the pump by removing its handle. People could no longer drink the infected water, helping to bring the outbreak to an end. Snow and other people interested in public health helped to convince the government that contaminated water pumped straight from the Thames was the likely culprit for the cholera outbreak, paving the way for cleaner water supplies and better sewerage. As people grew to understand the links between infectious disease, environmental factors, and immunology a little better, public health gradually started to improve.

Living conditions

The industrial revolution in England led to many advances in science, technology, and industry. British cities grew dramatically as farm workers were attracted to the idea of city living and easier work. However, this rapid migration from the country to the town took place without appropriate city planning—or, in some cases, without any planning at all. The massive rise in urban populations brought with it significant problems. Sanitation systems proved completely inadequate to deal with the amount of human waste now being produced by these larger populations, and raw sewage was often dumped into rivers and water supplies from overflowing sewers. This contaminated drinking water, and led to outbreaks of disease such as typhoid and cholera, as you saw above. Housing was often inadequate and hugely overcrowded, allowing diseases to spread easily from room to crowded room. Food in the cities was relatively expensive, and families often found themselves too poor to sustain themselves. Malnutrition was rife, bringing with it vulnerability to non-communicable diseases such as rickets and scurvy (both caused by insufficient vitamins in a poor diet), plus an increased susceptibility to infection caused by a weakened immune system.

Access to cleaner water supplies and a better understanding of causes of infectious disease helped people to avoid infection, but when disease occurred it was still largely untreatable. Government acts such as the Sanitary Act of 1866 made local authorities responsible for the removal of 'nuisances to public health', and for the removal or improvement of slum dwellings. In the early part of the twentieth century, vaccination against diseases such as tuberculosis, whooping cough, diphtheria, and tetanus all became available, and vaccination began to spread around the world as an effective means of disease prevention. The discipline of public health grew to governmental responsibility for maternal and child welfare, health visiting, school medicine, sexually transmitted infections, and learning difficulties. From the 1850s to 1930, the average life expectancy rose from forty years to 60.8 years of age—largely brought about by improvements in living conditions and public health reforms.

The National Health Service

The National Health Service (NHS) was established in the United Kingdom in 1948, aiming to provide free healthcare coverage to all. Local government health departments initially retained a separate identity, but roles were quickly created in the new health service which overlapped with these more traditional offices. As effective treatment for infectious diseases increased during the second half of the twentieth century, the role of public health officials began to take on a more preventative role, dealing with diseases of lifestyle and ageing, such as heart disease and cancer. Infectious diseases remained important, however, and the science of epidemiology remained as relevant as ever in plotting outbreaks of disease and identifying possible causes.

Today, most epidemiological studies and tracking of infectious diseases occurs through a body known as Public Health England (PHE). Many diseases (such as chest infections, colds, and other less serious infections) do not require monitoring or tracking. However, Public Health England maintains a list of so-called 'notifiable diseases'. These are infectious diseases which can make people seriously ill, which can spread quickly from person to person, or which are commonly caused by environmental problems (such as contaminated food or water supplies). If a doctor, nurse, or other healthcare professional sees a patient with one of these diseases, they must report it to their local PHE office, who will investigate the case further. If the report is one of several in an area, PHE may declare an outbreak of disease. The organisation will then undertake contact tracing, attempting to identify where the disease has been spreading from and ensuring that all cases are appropriately treated and followed up. If the outbreak comes from an environmental source, PHE may require evidence that the source is no longer a threat—for example, that a particular batch of foodstuff which has been found to be contaminated with bacteria has all been recalled. PHE also collects data on all the outbreaks which it investigates. This allows it to build up a national picture of infectious diseases and determine any 'hot spots' of infectious disease outbreaks which require further investigation or treatment. This helps the efforts of local teams to have a national impact on disease prevention and control.

Worldwide, a similar approach is adopted. The WHO collects data from the public health organisations of many countries around the world, allowing it to examine patterns and trends in infectious disease worldwide. It employs teams of epidemiologists, statisticians, healthcare workers, and other specialists in public health. Ordinarily these individuals are based in WHO centres, but if an outbreak of disease appears sufficiently serious, they may be deployed to the area of the outbreak, to monitor, advise, treat, and contain the infection as far as possible. The WHO has worked to create a 'network of networks'—connecting local, regional, national, and international public health agencies and laboratories to allow rapid sharing of data on outbreaks. You can read more on the challenges facing modern public health organisations in Chapter 7.

Chapter summary

- Vaccinations work by introducing the body to a dead or weakened version of a pathogen. This allows the T and B memory cells of the adaptive immune system to store the 'memory' of these specific antigens, allowing the immune system to quickly respond if the host meets the same pathogen again in the future.
- Herd immunity helps to reduce the spread of infectious disease through a community—the more individuals who are immunised, the harder it becomes for a disease to spread to people who are still vulnerable to it.
- Basic public health measures—such as street cleaners and spittoons—have existed for a long time. It is only relatively recently that governments and health organisations have started to track the spread and cause of infectious diseases.
- Understanding the spread of diseases can help to identify their source, and potentially help public health scientists to identify a removable or treatable cause for an outbreak.
- The WHO aims to link local, regional, national, and international healthcare organisations and laboratories into a worldwide surveillance network for infectious diseases, aiming to identify and treat large-scale outbreaks before they become pandemics.

Further reading

1. 'Vaccines and Autism—A Tale of Shifting Hypotheses'. https://academic. oup.com/cid/article/48/4/456/284219/Vaccines-and-Autism-A-Tale-of-Shifting-Hypotheses

   A paper looking at the arguments commonly put forward against vaccinating children, covering the MMR in greater detail, along with other common questions and theories.

2. https://www.gov.uk/government/publications/the-complete-routine-immunisation-schedule

   This link gives you access to the most recent comprehensive list of the current vaccination schedules in the UK. You will have had many of these vaccinations yourself—usually when you were far too young to remember them!

3. 'Public Health and English Local Government: Historical Perspectives on the Impact of "Returning Home"'. https://academic.oup.com/jpubhealth/article/36/4/546/1532200

   A paper discussing the historical elements of public health in the UK, along with modern challenges facing Public Health England.

4. 'Global Infectious Disease Surveillance'. http://www.who.int/mediacentre/factsheets/fs200/en/

   Information from the WHO on their role in disease surveillance, and their sources of information.

Discussion questions

6.1   Vaccination is a phenomenally powerful tool in disease prevention. Can you think of any reason why more diseases are not vaccinated against? What barriers might there be in developing an effective vaccine to a pathogen which we do not currently vaccinate against?

6.2   How can an organisation like the WHO effectively coordinate the epidemiological efforts of the entire world? What steps do you think need to be taken to prevent either over-reporting of diseases which are not concerning to the public, or under-reporting of potentially very serious outbreaks. What other factors do you think might affect this reporting from country to country?

7 GLOBAL OUTBREAKS, A POST-ANTIBIOTIC LANDSCAPE, AND THE EVOLUTIONARY ARMS RACE

Over the past six chapters, you have learned something not only about the pantheon of pathogens that attack our bodies every day, but also about the amazing advances in medicine over the centuries—to say nothing of the incredible defences that our own bodies are capable of mounting against infectious diseases. It would be easy to assume that we are growing ever closer to eliminating infectious disease entirely—surely global vaccination campaigns, effective antimicrobial treatments, and diligent public health surveillance will allow us to destroy our microbial invaders forever, if we can only contribute enough time and money to the cause? Unfortunately, the reality could not be further from the truth. Infectious diseases still run rampant throughout the world, and our arsenal is beginning to shrink, as antibiotic resistance spreads around the globe. As ancient explorers discovered new continents, they brought with them new diseases, often with devastating effects, as shown in Figure 7.1. Tracing and tracking outbreaks is now, in some ways, easier than ever before—but the ability to travel around the entire globe in a few hours also makes it so much easier to carry pathogens from one place to another. Ease of travel also makes it potentially very challenging to follow up infected individuals, and can make contact tracing a nightmare for epidemiologists and public health officials. Rather than rejoicing in the achievements of the past few centuries, should humanity be preparing for a future in which, once again, infectious diseases are often untraceable and untreatable?

Figure 7.1 The movement of diseases between continents is not new. When Europeans first travelled to the Americas, they took with them diseases such as smallpox and measles. The indigenous populations had no immunity to these new pathogens, and their communities were decimated by the incoming diseases.

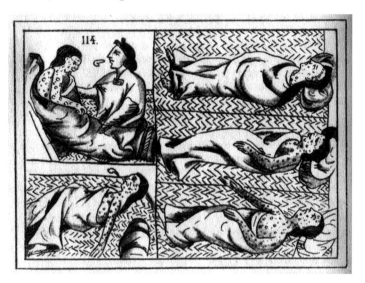

GL Archive/Alamy Stock Photo

The global village

Over the past hundred years or so, the world has become a much smaller place. Journeys which once took weeks or months overland or by ship can now be achieved in mere hours by aircraft. Travel has become cheaper, allowing more and more people to visit new places around the world (Figure 7.2). Alongside the movement of people, the movement of commercial products has also become much easier and lower cost, with many of the products we use every day coming from far-flung corners of the globe. This easy transport of people and foodstuffs also gives microbes a golden opportunity. People harbouring infectious diseases can often still spread the disease even if they have no symptoms of the disease at all. Other symptoms—such as coughing, sneezing, or diarrhoea—may not be severe enough to prevent a sufferer from travelling. Whilst waiting in an airport, an infected person is surrounded by thousands of others who may be travelling to multiple different destinations around the world. One cough or sneeze can spread millions of bacteria or viral particles through airborne or droplet spread, each potentially infecting somebody in the close vicinity of the initial 'patient zero'. After sitting in the airport for a while, the patient then spends up to twelve hours on an aircraft with hundreds of others, potentially infecting some of them with each cough or sneeze. In this way, it is possible for one person to infect many hundreds (or even thousands) of others—and for this to trigger a chain reaction of infection, as each of these people goes on to infect others around them.

Figure 7.2 Cheap flights have shrunk the world we live in, but they have also provided pathogens with a rapid route from continent to continent.

Nieuwland Photography/Shutterstock.com

This ease of transportation around the world presents a very real challenge for public health surveillance. By the time an outbreak of infectious disease has been noticed in one part of the world, some of those infected may have already travelled to an entirely different continent—often without even realising that they are harbouring an infection. In the event of a severe outbreak, how can public health officials track these individuals, let alone the many other people who may have been exposed to them whilst they travelled? Sometimes the infected traveller may be identified—if they become unwell enough to present to a healthcare professional and are diagnosed correctly, or if they die of the infection. However, sometimes outbreaks may only be picked up after they have begun, and contact tracing may only be feasible once the disease has been identified.

Political problems

As well as these physical challenges, organisations such as the World Health Organization (WHO) face significant political and financial challenges to successful public health endeavours. Even in developed countries with sophisticated healthcare systems, disease surveillance is not perfect, and public health measures are never foolproof. Cuts to public services may reduce staff numbers or resources to levels where effective surveillance becomes more challenging, and health scares such as the MMR controversy of the late twentieth century (discussed in Chapter 6) may reduce uptake of effective public health measures. In the developing world, public health officials may suffer from lack of resources and poor access to health care for large swathes of the population (meaning that many sick people never get to see a doctor, let alone receive a diagnosis), or local political upheaval. In areas of the world where civil unrest or war is ongoing, obtaining accurate epidemiological information is very difficult, and the movement of refugees and migrants can make tracking outbreaks almost impossible.

Local political and cultural factors may also make it difficult or dangerous for public health officials to perform their jobs, rendering public health campaigns extremely hazardous to implement. In Pakistan, for example, the polio vaccination campaign has met significant resistance, with health workers being executed by local Taliban groups, who are opposed to the vaccine. These groups claim that the vaccination programme is against their religious beliefs, and believe that it is part of a programme to make the local population infertile. As a result, Pakistan—in many ways a developed nation with much to offer—remains one of the few countries in the world where polio has not been eliminated. Even in large, economically developed countries, language and culture can create a significant barrier to effective worldwide disease surveillance. Case study 7.1 shows some of the challenges that can face public health officials when trying to track the first stages of a global outbreak.

Case study 7.1
SARS: tracking a pandemic

The WHO aims to track and study disease worldwide, relying on local laboratories and authorities to accurately report outbreaks of infection. When this global network of surveillance is successful, it is a phenomenally useful tool for gathering data on new and potential outbreaks of disease. However, if a disease is not reported due to the inability or unwillingness of local authorities, the consequences can be deadly. When a new infectious disease appeared in China in 2002, a range of factors led to an outbreak which killed over 700 people and infected many more around the world (Figure A).

In November 2002, a farmer from Guangdong province in China presented to his local hospital with symptoms similar to a severe case of influenza. Suffering from fevers, breathlessness, cough, and malaise, he was admitted to the hospital and died shortly afterwards, with no definitive cause of death being identified. He was to be the first of many. Soon afterwards, other residents of Guangdong province started to fall ill with similar symptoms, with at least several hundred people infected, and a number of deaths. At this stage, the Chinese Government had identified the outbreak, but discouraged the Chinese press from reporting on it, and suppressed information about the outbreak everywhere except Guangdong province.

In late November 2002, Canadian public health officials picked up reports of a 'flu outbreak' in Guangdong province. Analysing reports from China, they determined that an epidemic of a new disease was possible, and sent the information on to the WHO. Unfortunately, due to language barriers and issues translating the initial reports, an English report into the outbreak was not published until January 2003. In the meantime, the WHO requested further information from the Chinese authorities, but received little cooperation. In February 2003, a doctor from Guangdong province named Liu Jianlun travelled to Hong Kong to attend a wedding. He had treated some of the patients presenting with this new flu-like illness, and had started to suffer from a fever and slight cough whilst travelling to Hong Kong. He was initially well enough to travel around Hong Kong and engage in some sightseeing, but rapidly became more unwell and was admitted to hospital on 22 February. From that point on, his condition worsened. He was admitted to intensive care, but sadly died on 4 March.

Figure A This map shows the countries affected by SARS in the 2002-4 outbreak, along with the numbers of infected individuals, and deaths caused by the disease.

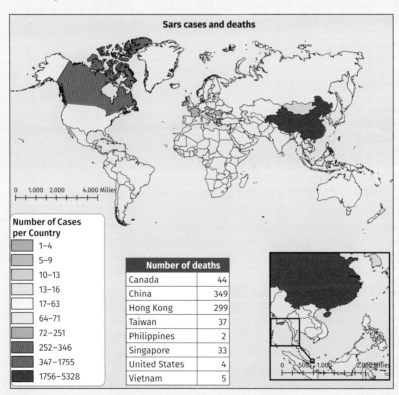

Sars cases and deaths

0   1.000 2.000   4.000 Milies

Number of Cases per Country
- 1–4
- 5–9
- 10–13
- 13–16
- 17–63
- 64–71
- 72–251
- 252–346
- 347–1755
- 1756–5328

Number of deaths	
Canada	44
China	349
Hong Kong	299
Taiwan	37
Philippines	2
Singapore	33
United States	4
Vietnam	5

0   500 1.000   2.000 Milies

In the meantime, a Chinese-American businessman named Johnny Chen, who had been staying in a hotel room across the hallway from the Chinese doctor, travelled to Hanoi in Vietnam. After his arrival, he fell ill with the same symptoms that other victims of the new illness had experienced—high fevers, malaise, and difficulty breathing. He was rushed into the French Hospital in Hanoi, where an Italian microbiologist named Carl Urbani formed part of the team treating him. Dr Urbani recognised that the symptoms that Mr Chen had presented with could represent the beginnings of an epidemic—although he initially believed the disease to be a form of avian influenza. He reported his concerns to the WHO before travelling to a medical conference in Bangkok, Thailand. He began feeling feverish on the flight, and asked a colleague who met him at the airport to call an ambulance. He was admitted to hospital in Bangkok, but his condition also deteriorated rapidly, and he died after a short stay in intensive care. Other members of the medical team who had cared for Mr Chen also fell ill with the same disease, and the WHO issued a global alert regarding a new infectious disease of unknown origin, which had infected people in both Hong Kong and Vietnam.

The disease—now given the name SARS, for Severe Acute Respiratory Syndrome—began to spread rapidly, appearing in Singapore and Canada. In April, the Chinese government began officially reporting on the infection, possibly prompted by increased international scrutiny of the outbreak after the death of an American teacher, James Salisbury. The cause of the infection was soon identified, firstly by a laboratory in British Columbia, and then by several other laboratories worldwide. The virus appeared to be in the corona-virus family, and was named 'the SARS virus'. A major change in Chinese policy appears to have occurred at this stage, with the Chinese Premier of the time, Wen Jiabao, stating that there would be severe consequences for public health officials not reporting outbreaks in a timely and accurate manner. Following this announcement, information from the Chinese public health authorities appeared to demonstrate that cases had been significantly under-reported on the Chinese mainland, an oversight which was put down to 'bureaucratic ineptitude' by Chinese authorities. This marked a significant turnaround from earlier stages of the outbreak, when WHO officials were not permitted to travel to Guangdong province to investigate the initial outbreak, and reports of outbreaks through China had been suppressed.

Quarantine protocols and travel restrictions were rapidly enacted by governments in affected countries, and by the end of May 2003 the number of cases worldwide had started to drop, with Beijing, Taiwan, and Toronto (Figure B) being cited as the only remaining areas of infection. The WHO declared the outbreak to be contained by July 2003, although isolated outbreaks in small numbers of people (often scientists working with the virus in laboratory settings) were reported until 2004. The disease had been successfully contained after infecting at least eight thousand people and killing 774 people in thirty-four countries worldwide. The virus was subsequently found to have a probable reservoir in bats (which are asymptomatic despite carrying the virus), and to have spread to humans either directly or through other mammals such as palm civets, which are frequently sold in Chinese markets.

Figure B This graph shows the two phases of the SARS outbreak in Toronto, Canada, in 2003. Patient zero in Canada was a seventy-eight-year-old woman who stayed in the same Hong Kong hotel as Dr Liu Jianlun. Around 400 people became ill and forty-four died of SARS in this one, rapidly controlled outbreak.

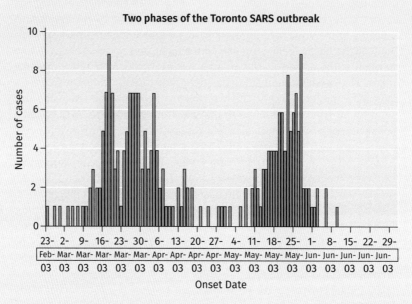

Canadian Environmental Health Atlas

? Pause for thought

After the outbreak of a new disease, reflection on the national and international response is obviously crucial to improving reactions to future outbreaks. Political and cultural reticence to share this information can make future planning quite challenging. Can you think of any ways in which a global infectious disease surveillance network could be made more effective? Do individual nations have a right to privacy when it comes to infectious disease, or should disclosure be mandatory? What other challenges can you think of when it comes to monitoring new and rapidly spreading outbreaks of disease?

The impact of time and place

The severity and impact of an infectious disease isn't only about the virulence of the pathogen. The time and place of an outbreak of disease can also have an impact. Influenza in the UK kills many more people in the winter than it does in the summer, in part because winter conditions mean at-risk groups such as the elderly are even more vulnerable due to the cold weather and an over-stretched health service. Sometimes local customs can help to spread infections. In late 2013, a village called Meliandou in Guinea, West Africa, became the epicentre of a deadly outbreak. A two-year-old child was the first to fall ill, with high fevers, diarrhoea, and vomiting. The disease quickly spread through the village, and then on to other local areas in the prefecture of Gueckedou. This region of Guinea was an important local trading hub, and people from Guinea, Sierra Leone, and Liberia all travelled to the area to trade. The two-year-old in Meliandou died in December 2013; by March 2014, the first case had been reported in Liberia. In May 2014, Sierra Leone had confirmed that it, too, had seen outbreaks of the infection. The WHO would later confirm the outbreak to be Ebola—a severe and deadly viral infection which causes high fevers, vomiting, diarrhoea, and profuse bleeding from all orifices. It is often fatal, and the virus is present in the victim's bodily fluids—blood, mucus, tears, and even semen.

Once a patient has started to experience the symptoms of Ebola, they are highly infectious to those around them, even after death. Local funeral customs often called for members of the family to wash and clean the corpse before burial. With little formal healthcare infrastructure, people often relied on faith healers. Unless adequate protective measures are taken, those caring for Ebola patients can easily become infected themselves. Medical teams looking after Ebola patients often use full body personal protective equipment (PPE)—covering their eyes, nose, mouth, hands, and body. If possible, the patient should be nursed in isolation, away from other patients—even ones who are also suffering from Ebola. There are no antivirals which can treat the ebolavirus, and there is currently no effective vaccine. Supporting a sufferer with intravenous fluids, rest, painkillers, and intensive care are the only options available to hospital staff—and the hospitals rapidly filled up. As a result, many sufferers were nursed at home by their friends and family. Sometimes, the patients' own immune systems would prove strong enough to fight off the infection, but sadly many could not overcome the virus, and subsequently died.

The WHO and other international medical organisations, in addition to local public health organisations, implemented infection control measures where they could. Hospital staff were equipped with PPE, the homes of infected people were disinfected with chlorine, and patients who were confirmed to have Ebola were isolated where it was possible to do so. Movement across country borders was restricted, and health authorities around the world were briefed about dealing with people returning from the affected countries. The spread of disease was largely contained to the local area, with a few isolated cases elsewhere in the world, almost all of which were from people who had travelled to the countries at the centre of the outbreak.

However, there was considerable opposition in the local population to some of the public health measures imposed. Many people had little education, and did not understand the causes of infectious disease. Some believed that the infection was a punishment for the sufferers. Others claimed that the healthcare workers brought it with them, to infect healthy Africans. People were understandably upset when told they could not have traditional funerals. These rituals had been practised for hundreds of years, and were the last opportunity a grieving family had to say goodbye. Faced with these challenges, public health officials had to try to engage with local people in a way they could understand, and to be culturally sensitive to their needs.

Posters were printed, explaining how people could prevent infection. Healthcare workers attempted to explain the origins of the virus to local

Figure 7.3 Posters like this one were used to help educate the local people as to the causes and symptoms of Ebola, as well as to dispel common myths about the disease.

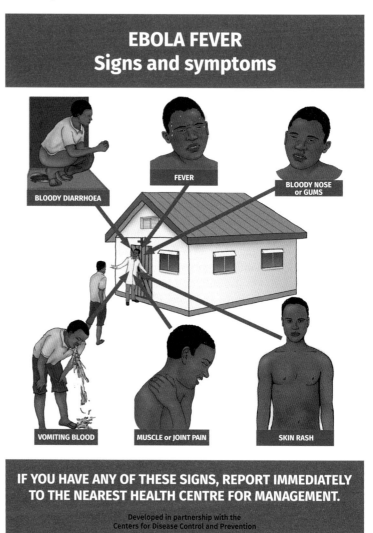

CDC

populations, including the reasons why infection control measures were important. You can see an example of local education posters in Figure 7.3. Gradually, the Ebola outbreak began to die out—but it had infected over 28,000 people, and claimed over 11,000 lives. The WHO admits that the total number of those infected or killed by Ebola is likely to be much higher, but that limitations in data collection in the affected countries means we may never know the true number of people affected. From this most recent outbreak, it is clear that fighting a modern pandemic cannot simply be accomplished with scientific infection control methods—although these play a crucial part. Managing the demands and expectations of the populace is critical, alongside education about the disease and necessary measures to control it.

The evolutionary arms race

The human race has taken many tens of thousands of years to reach our current state. In that time, we have become quite good at trying to avoid infectious disease—with only those humans who have survived life's challenges tending to be the ones who successfully went on to reproduce. Through trial and error, we have learned as a species that eating non-spoiled food, drinking clean water, and avoiding physical injuries can help us to avoid infections, and lead longer, healthier lives. However, pathogens have not sat idly by, either. Some have evolved alongside us in very clever ways, ensuring as far as possible that they, too, are able to survive and reproduce. Take the Guinea worm, for example. This is a nematode, which lives its first larval stages in fresh water. They take up residence in the body of copepods, small crustaceans which are found all over the world, where they continue to develop. Ultimately, their goal is to be ingested by a person who, taking a drink of water, also swallows the copepod and its parasitic load. Look at Figure 7.4—it shows the larval stages of the worm.

Figure 7.4 The early stages of the Guinea worm life cycle. In stage 6 of the cycle, you can see the larvae infesting a freshwater copepod.

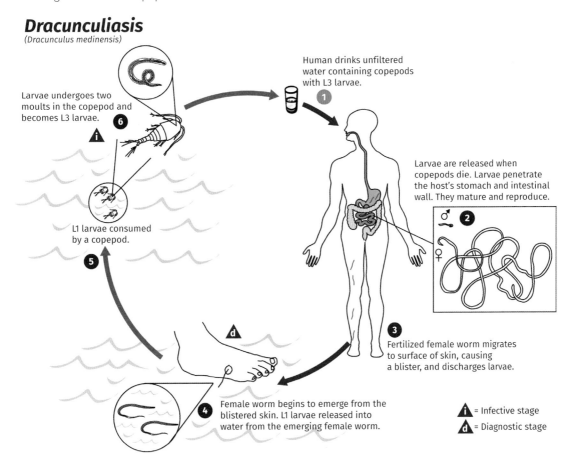

Dracunculiasis
(*Dracunculus medinensis*)

Larvae undergoes two moults in the copepod and becomes L3 larvae. **6** **i**

Human drinks unfiltered water containing copepods with L3 larvae. **1**

L1 larvae consumed by a copepod. **5**

Larvae are released when copepods die. Larvae penetrate the host's stomach and intestinal wall. They mature and reproduce. ♂ **2** ♀

Fertilized female worm migrates to surface of skin, causing a blister, and discharges larvae. **3**

Female worm begins to emerge from the blistered skin. L1 larvae released into water from the emerging female worm. **4** **d**

i = Infective stage
d = Diagnostic stage

Public Health Image Library (PHIL)/CDC/Alexander J. da Silva, PhD/Melanie Moser

Once a female worm has mated she migrates to the tissue just under her host's skin, usually in the feet or lower legs. After twelve months the eggs hatch inside her, filling her body with millions of first-stage (L1) larvae. Now, she needs to access water for her offspring. They can only survive in a freshwater environment, and they must find another copepod in which to continue their life cycle, otherwise they will die. The worm has evolved an ingenious way of forcing a human to seek out water.

The female worm releases a few of her larvae into the skin, resulting in an innate immune response with inflammation, swelling, and a painful blister in the skin. The blister causes an intense burning sensation in the skin around the lesion. In an attempt to relieve this, the host will often plunge the affected limb into cool, soothing water. A section of the female ruptures, expelling millions of her offspring into the water, where they search for the copepod which forms the host for the next stage of their life cycle. This is repeated many times, over several weeks. Eventually, the body of the female is completely removed from the host, when the skin often heals quite quickly. However, victims will often suffer from chronic pain, or even physical deformity, after the infection. Look at Figure 7.5—it shows a guinea worm emerging from the skin of an infected person—in this case, on the abdomen of a child.

Guinea worm infestation has become increasingly rare, thanks to the work of the Carter Foundation—a humanitarian organisation dedicated to the eradication of the worm. Since 1986, they have reduced the number of guinea worm infestations from 3.5 million to just twenty-five cases in

Figure 7.5 The traditional treatment when the worm emerges is to wind it around a piece of gauze or thin piece of wood, and to very slowly twist the worm out of the body. This painful process can take many weeks, leaving plenty of opportunities for secondary bacterial infections to take hold in the open wound.

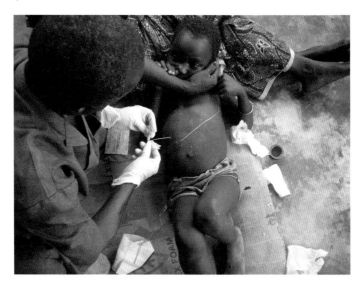

Photo: Mary Glenshaw, Ph.D., M.P.H.; OTR/L; OD/OWCD/CDD/EFAB. Public Health Image Library (PHIL)/CDC/CDC Connects

total worldwide! Their work is largely educational, persuading local people to filter water before drinking it, and preventing anyone with an emerging worm from entering local water sources.

The guinea worm is not the only pathogen that manipulates our behaviour. Even simple cold viruses spread by utilising our coughing and sneezing mechanisms. There are other, more subtle ways in which infectious diseases can change the way we act and perhaps even the way we think, as you can see in Case study 7.2.

Case study 7.2
Toxoplasma: mind control and methylation

Toxoplasma gondii is a harmless-looking little organism. Only visible under the microscope, it has no teeth, no hooks, no barbs, and is similar to many other protozoa. In its ordinary life cycle, it infects rats and other rodents, before reproducing within the digestive tracts of cats (Figure A). An infected rodent is consumed by a cat—either wild or domestic— which is its **definitive host**. Once inside the intestines of the cat, thousands of hardy *T. gondii* **oocysts** are passed out every time it defecates. These oocysts may lie dormant for many

Figure A Domestic cats have been important human companion animals for thousands of years—but we may be sharing our homes with more than just the family pet.

Ann Fullick

Figure B *Toxoplasma*—commonly found in cats, rats . . . and people.

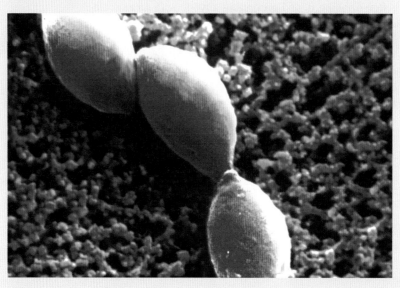

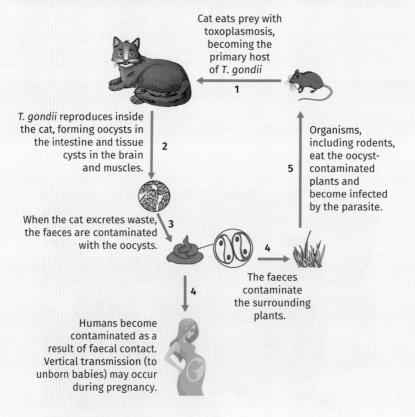

Cat eats prey with toxoplasmosis, becoming the primary host of *T. gondii*
1

T. gondii reproduces inside the cat, forming oocysts in the intestine and tissue cysts in the brain and muscles.
2

Organisms, including rodents, eat the oocyst-contaminated plants and become infected by the parasite.
5

When the cat excretes waste, the faeces are contaminated with the oocysts.
3

4

The faeces contaminate the surrounding plants.

4

Humans become contaminated as a result of faecal contact. Vertical transmission (to unborn babies) may occur during pregnancy.

months, but once they are ingested by a warm-blooded animal, the oocyst dissolves and the life cycle continues, as you can see in Figure B.

Once *T. gondii* has successfully colonised a rodent, it has a problem: in order to reproduce, the parasite needs to be inside a cat—but rats and mice instinctively avoid cats. Amazingly, this unassuming little protozoan has evolved an ingenious method of encouraging its host to become a meal for a cat. It brings about epigenetic changes to DNA in the **amygdala** of its host's brain, remodelling it and dramatically reducing the natural predator aversion of its rat host. Studies show that rats and mice infected with *T. gondii* do not avoid areas contaminated with cat urine, whereas their uninfected counterparts stay well away. In addition to this, infected subjects showed poorer motor skills and coordination, making them easier for a cat to catch. In short, toxoplasma is a mind-controlling parasite. By manipulating the behaviour of its host, *T. gondii* improves the likelihood that it will be transmitted back to its definitive host, and therefore increases its probability of reproducing.

But what happens if toxoplasma finds itself in the wrong host? Although it aims to infect rodents, the oocysts of *T. gondii* can infect most mammals, including humans. In fact, it has been estimated that up to 50% of humans on the planet have been exposed to (and may be chronically infected by) the little protozoan. Ordinarily, infection with *T. gondii* causes no symptoms at all. In vulnerable people—particularly the immunosuppressed—it may cause symptoms similar to the 'flu, along with swollen lymph nodes. Occasionally it may also infect the eyes (causing sight loss) or the lungs. If a woman becomes infected with toxoplasma whilst pregnant, she can pass the infection on to her unborn baby. This can result in a condition called **congenital toxoplasmosis**, which may lead to conditions such as epilepsy, **hydrocephalus**, jaundice, brain damage, learning difficulties, or **cerebral palsy** in the baby.

Can toxoplasma manipulate *our* behaviour? There is conflicting evidence. Some research suggests that men infected with *T. gondii* may be more likely to take risks, ignore rules, and be more suspicious and jealous than their non-infected counterparts. In the same studies, infected women were more likely to be more outgoing, warm-hearted, and conscientious. Other studies have linked *T. gondii* infection with poorer motor skills, suggesting that infected individuals are more likely to be involved in motor vehicle accidents. There has also been some evidence to suggest that toxoplasma infection could also increase the risk of some mental illnesses such as **schizophrenia** or **bipolar disorder**. These links are very difficult to prove, however, and other studies have shown no links between *T. gondii* infection and mental illness. For now, the jury is out.

❓ Pause for thought

Think about how you could examine the link between T. gondii infection and behavioural changes. What sort of study might be best-suited to examine this? How would you approach recruiting participants? What might the limitations be?

Antibiotic resistance—life in a post-antibiotic landscape

You have already learned how quickly bacteria adapt to their surroundings. By transmitting plasmids to one another, they can transfer DNA horizontally and confer evolutionary advantages to their peers as well as their offspring. Knowing this, it should come as no great surprise to learn that bacteria are very good at becoming resistant to the antibiotics we use to treat them. Even Alexander Fleming, discoverer of penicillin, warned that over-use of antibiotics could lead to humanity essentially 'selectively breeding' drug-resistant bacteria. At the time, the scientific establishment paid little heed to his warnings—but they seem eerily prescient today.

Antibiotic resistance can occur in one of two ways. A bacterium may have what is known as intrinsic resistance, which occurs when the natural characteristics of the organism make it resistant to certain types of antibiotic. This is often, at least in part, due to the fact that many microbes have been using antibiotics against one another for millions of years. The ability to secrete an antibiotic obviously gives a bacterium a distinct advantage in its natural environment—it is able to kill off its competitors and use the space and resources that they leave behind. In order for a bacterium to effectively use an antibiotic, however, it must also be resistant to it, otherwise it would risk killing itself off as well as its rivals! Over millions of years, various different bacteria (many of them completely harmless to humans) have evolved the ability to secrete complex antibiotic compounds. Other mutations have helped bacteria develop resistance to these compounds, creating an endless cycle of offence and defence among bacterial colonies. This intrinsic resistance has been demonstrated in many different environments—even amongst bacteria which have never been exposed to antibiotics manufactured by people. Widespread antibiotic resistance was recently discovered in bacteria living deep underground, in a cave system which had been isolated from the rest of the world for four million years! Even more remarkably, some of the bacteria found within this cave system demonstrated resistance to synthetic antibiotics which did not exist on Earth until the twentieth century. Antibiotic resistance can therefore be said to be widespread throughout the natural world, even without human interference.

Acquired resistance refers to the ability for bacteria to develop resistance to the common synthetic antibiotics used against them. This tends to occur in one of two ways. Rarely, the presence of an antibiotic may induce resistance mechanisms in the bacteria exposed to it. More commonly, however, the introduction of an antibiotic leads to a good example of 'survival of the fittest' amongst the bacteria. Through the duration of a course of antibiotics, the bacteria which are most susceptible to the antibiotic are rapidly killed or prevented from reproducing. Any bacteria which have mutations protecting them against the action of the antibiotic are therefore free to reproduce, taking advantage of the space and resources not being used by the bacteria which have been destroyed by the antibiotic. Look at Figure 7.6—it gives you a better idea of how this works.

There are a number of different factors at play in the rise of acquired antibiotic resistance. The first, of course, is the use of antibiotics in human

Figure 7.6 In this example, bacteria which do not have any resistance to the antibiotic are quickly killed off, leaving bacteria which have one or more positive mutations which allow them to resist destruction. These resistant bacteria can then reproduce in greater numbers, as they have more space and resources.

A group of bacteria, including some which are naturally resistant to an antibiotic,

are exposed to antibiotics. Most of the bacteria without a degree of resistance are killed off

The resistant bacteria multiply and become more common.

Eventually, the entire group of bacteria is made up of those that are resistant to the antibiotic

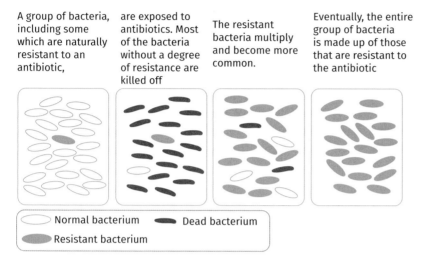

○ Normal bacterium   ━ Dead bacterium
● Resistant bacterium

medicine. Since their invention, humanity has used many millions of kilograms of antibiotics to treat bacterial infections, often with great success. However, as we have learned above, bacteria are extremely adept at evolving mechanisms of resistance to the antibiotics used against them. This became evident very rapidly, with resistant cases of *Staphylococcus* infections in American hospitals rising from 14% in 1946 to 59% in 1948, for example. Before long, other drug-resistant organisms had begun to emerge, but the medical establishment maintained that new drugs would be developed which would be able to successfully treat these resistant bacteria. Initially, they were right—many more antibiotics were developed between the 1950s and the 1970s, but antibiotic resistance followed closely behind. The supply of new antibiotics began to dwindle after the 1970s. Gradually, microbiologists began to see the rise of so-called 'superbugs'—bacterial strains resistant to multiple different antibiotics, making them even more challenging to treat.

Antibiotics and animals

In tandem with the use of antibiotics in medicine came the use of antibiotics in agriculture and veterinary medicine. Throughout the world, many animals used as livestock are given antibiotics, even when they are not unwell. Often mixed in with their feed, this treatment aims to keep them healthy and to improve their growth, reducing the number of animals lost to disease, and therefore increasing profits when the animals are sold. Although the UK does not allow antibiotics to be used in this way, it is normal practice in countries such as the United States. This indiscriminate use of antibiotics allows any bacteria which colonise the healthy animal to quickly develop antibiotic resistance, through the mechanism described in Figure 7.6. When the animal passes these bacteria into the environment around it, the antibiotic-resistant bacteria can find their way into other animals, including humans, where they can cause serious illness. The spread of these resistant bacteria may be considerably wider than we had previously thought: even wildlife in the Arctic Circle has been shown to demonstrate alarmingly high levels of single- and multi-drug-resistant bacteria such as *E. coli.*

What can we do?

Although we have understood the mechanisms of antibiotic resistance for many years, it is only relatively recently that we have been forced to take action. The WHO now considers antibiotic resistance to be one of the greatest threats currently facing humanity. They state that antibiotic resistance is 'threatening our ability to treat common infectious diseases, resulting in prolonged illness, disability, and death. Without effective antimicrobials for prevention and treatment of infections, medical procedures such as organ transplantation, cancer chemotherapy, diabetes management, and major surgery become very high risk.' It is becoming very clear that, without effective alternatives to antibiotic use, humanity is likely to find itself once more at the mercy of infectious disease. So, what are our options?

Reduce antibiotic use

Firstly, it is hugely important to reduce the amount of antibiotics used in both humans and animals. If we can achieve this, then we may be able to slow the rate of antibiotic resistance. Ultimately, if we can stop using certain classes of antibiotics for long enough, bacteria may become sensitive to them once again. To achieve this, people need to learn appropriate antibiotic stewardship—prescribing antibiotics only when and where they are strictly needed. Although this task sounds simple, there are many difficulties. For example, at present, if some of the birds in a commercial flock of chickens fall ill, the vet will often treat the entire flock to ensure eradication of the disease. However, research from as early as the 1970s has shown that a week of antibiotics is enough for resistant bacteria to show up in the faeces of the chickens. Similar difficulties occur throughout the veterinary world, and also apply to human medicine. Differentiating between a viral and a bacterial infection isn't always easy—for example, the symptoms of a viral and a bacterial sore throat may be very similar indeed—and although healthcare professionals try to minimise prescription of antibiotics for conditions where they may not be needed, this isn't always possible. People suffering from chronic health conditions are much more susceptible to infections, and are often treated with antibiotics very early on in the course of their infection, regardless of the pathogen. In many countries around the world, it is also possible to buy antibiotics over the counter at the pharmacy, rather than requiring them on prescription. This results in people taking unnecessary antibiotics for viral illnesses (against which they are useless).

Reducing the use of antibiotics in humans and animals requires an extensive education campaign on the part of national and international public health organisations. The general public needs to be educated on the appropriate use of antibiotics, and healthcare professionals need to be encouraged not to prescribe antibiotics if they do not feel that they are necessary. This can be quite challenging, particularly for patients who do not understand the difference between a bacterial and a viral infection, and who may feel that they 'need' an antibiotic to make them feel better. You may have seen public health campaigns, such as the one shown in Figure 7.7, which attempt to educate people to the appropriate uses of antibiotics, whereas Figure 7.8 discusses sensible antibiotic prescribing with healthcare professionals.

Figure 7.7 This poster forms part of a campaign to encourage people to think about the appropriate use of antibiotics—what they will and won't work to treat.

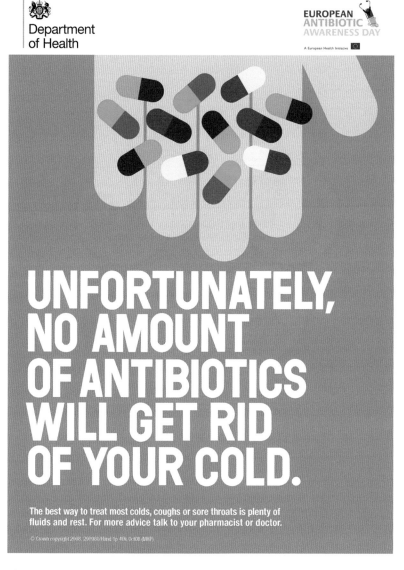

The Department of Health and Social Care

Figure 7.8 This poster aims to remind healthcare professionals about responsible prescribing, encouraging bacterial cultures, use of appropriate agents via the appropriate route, and an appropriate length of course.

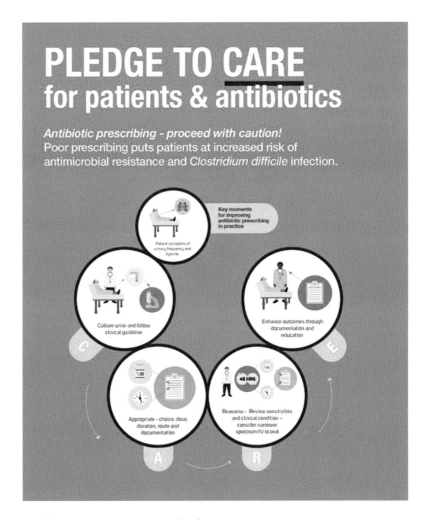

Healthcare Improvement Scotland

Of course, effective antibiotic stewardship comes with its own risks. A good example of this is the culturing of samples to check the strain of bacteria and which antibiotics it is sensitive to. Often, the sample can be taken very quickly (either via a swab, urine sample, or blood sample, depending on what needs to be cultured), but it sometimes takes a few days (or even longer) before a result is available—and in the meantime, the patient may have deteriorated, becoming more unwell. In many cases, an antibiotic needs to be started before the results of the culture are known—even if this ultimately turns out to be an antibiotic which the bacteria is resistant to, necessitating a change in treatment.

Genomic sequencing is one area which has huge potential to prevent inappropriate prescribing. Whole genome sequencing (WGS) is a potentially rapid way of telling us what pathogen is involved in an infection. The original sample does not need to undergo specific culturing, and any genes coding for specific virulence—such as the ability to produce particular exo- or endotoxins—or antibiotic resistance, can be rapidly identified, allowing microbiologists to obtain a result within hours, rather than days. Because the clinical application of this technology is mainly to identify the strain of bacteria along with any virulence factors and resistance it may have, it is not always necessary to completely sequence the genome, speeding up delivery of the results even further. Ultimately, it may become possible for genomic sequencing to be performed within minutes, opening up the possibility of 'point-of-care' testing by healthcare workers—giving doctors the opportunity to work out whether an infection is bacterial or viral whilst examining their patient, and further improving appropriate antibiotic prescribing.

The importance of public health

Good public health measures are also tremendously important to reduce the number of individuals affected by an individual disease. Vaccination programmes are effective, but only with a high enough uptake to confer herd immunity (as you have already seen in Chapter 6)—and, of course, not all infectious diseases can be vaccinated against. Simple measures—such as covering the mouth and nose when coughing and sneezing, and observing good hand hygiene, both by the public and by healthcare workers—can be surprisingly effective at combating the spread of disease either via the airborne or faecal–oral routes. In a hospital environment, a patient with a potentially infectious disease may be nursed in a room away from other patients, and aprons, masks, and disposable gloves should be used when interacting with them.

However, even with the most diligent and appropriate use of antibiotics, combined with flawless public health measures, human beings are still going to fall ill with infectious disease. There's simply no way to completely prevent it. So, if antibiotics are becoming less effective, what options do we have left? The first suggestion to this question is usually 'develop more antibiotics', but this is proving very difficult. New antibiotics are difficult to synthesise, expensive to develop and test, and may not even prove safe for human use after undergoing years of laboratory work and millions of pounds in research. We should not rely on the ability to discover new antibiotics forever—and new antibiotics are likely to quickly run up against the same problems of resistance as the current drugs. Sometimes, old antibiotics may be repurposed—and sometimes it may also be possible to alter the structure of an existing antibiotic to thwart the resistance mechanisms of particular bacteria. But we need new ideas.

Alternatives—thinking outside the box

An alternative to simply developing or adapting antibiotics is to consider using other infectious diseases to combat bacteria. Bacteriophages are viruses which specifically target bacteria, infecting them in the same way that some viruses infect human cells. Bacteriophages exist in the natural world as obligate parasites of bacteria, infiltrating the prokaryotic cell in a very similar way to viruses affecting humans. Once inside the bacterium, the virus hijacks its organelles to reproduce, before triggering lysis of the bacteria, destroying it. However, there is another possible use for bacteriophages. Instead of causing outright destruction of the bacterial cell, certain types of bacteriophage simply insert their own genetic code into the DNA of the bacteria, forcing their own genes to be expressed within the bacterium itself. This feature allows researchers to insert genes re-introducing vulnerability to antibiotics into colonies of resistant bacteria, removing the bacteria's acquired resistance, and allowing it to be treated with current antibiotic therapy once more. The advantages of bacteriophages are numerous:

- they only affect the cells of the bacteria which they parasitise, and so cause minimal disruption to the human immune response;
- they destroy individual bacteria or make them vulnerable to antibiotics;
- they are capable of self-replication, allowing a small number of bacteriophages to potentially treat a significantly greater number of bacteria.

Some scientific trials have shown that they are effective for treating resistant bacterial infections, but much more evidence is needed before they become a commonly used treatment in the Western world—although they are used quite extensively in Russia and the former states of the Soviet Union.

There are a number of potential problems with bacteriophages. Firstly, we don't fully understand their impact on the human body. In theory they should not be able to cause disease in a human host as they do not infect human cells—but they may still be able to trigger an immune response, as they will be identified as 'non-self' by the immune system. There are also some worries around how they might react in an immunocompromised person. Another concern is the adaptability of both the bacteriophage and the bacteria it is designed to infect. If the bacteriophage mutates, allowing it to infect human cells, then it could go on to cause disease in people. It also seems to be possible for bacteria to develop resistance to bacteriophages, in much the same way as they do to antibiotics. Many bacteriophages rely on a single glycoprotein to act as their entry point into a bacterial cell—and if the target bacterium evolves to remove this from its cell wall or membrane, the bacteriophage cannot gain entry. Of course, another disadvantage to bacteriophages is that each virus can only infect a limited number of different bacteria, and may even specialise in just one species of bacterium. Therefore, it seems unlikely that we will discover any 'broad-spectrum' bacteriophages. The potential use of bacteriophage therapy makes pathogen identification even more crucial. It is, of course, possible that rapid WGS could provide a partial answer to this problem. Ultimately, we may be able to 'tailor' specific treatments for people with resistant bacterial infections—performing genomic sequencing to identify the strain of bacteria infecting them along with any resistant properties, then custom-building a bacteriophage to either target the bacterium directly or to remove its resistant properties, opening it up to treatment with traditional antibiotic therapy.

A universal answer to the problem of antibiotic resistance is a long way off. Any new treatment that is developed needs to undergo many years of testing in a laboratory environment before it ever sees human trials—the average time for an antibiotic to complete all clinical trials and reach a stage where it can be prescribed is between ten and fifteen years! In the meantime, humanity continues to face the increasing threat of infectious disease with greater understanding than ever before, and an increasing toolkit of infection control measures. In a world with an ever-expanding population and increasing large-scale travel, will it be enough to hold back the next global pandemic? Only time will tell.

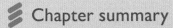 Chapter summary

- Improving global transport links, as well as the demand for food from all over the world, makes it easier than ever before for diseases to spread over great distances.
- Worldwide organisations such as the WHO aim to coordinate the efforts of local, national, and international public health organisations to track, treat, and prevent infectious disease.
- These efforts can sometimes be hindered by cultural or political boundaries, and sometimes disease can even be inadvertently spread by local customs or superstitions.
- Infectious diseases have evolved alongside humanity, and are constantly involved in an 'evolutionary arms race'—developing ways to subvert our defences, or even manipulate our behaviour for their own good.
- Bacteria can rapidly evolve, sharing genetic traits to increase their resistance to antibiotics.
- New antibiotics are both difficult and expensive to discover, and many infections are now developing resistance to multiple different antibiotics.
- It is possible that we will end up in a world where most of our antibiotics are no longer effective against many different bacteria.
- To avoid this, we must both try to reduce our use of antibiotics, and look for new or alternative ways of treating bacterial infections.
- Rapidly sequencing the genome of bacteria could allow us to quickly determine the cause of an infection, along with any resistances it may have. This could allow us to treat infections more accurately, using antibiotics in a more appropriate fashion.

- Good public health measures can help to reduce the number of infections that the population suffers from.
- Bacteriophages are viruses which infect bacteria, either killing them or altering their ability to resist antibiotic therapy. Although they are not widely used or well understood at present, they may also become an important alternative to antibiotics.

Further reading

1. BBC, 'Ebola—Mapping the Outbreak', January 2016. **http://www.bbc. co.uk/news/world-africa-28755033**.

   A BBC news review of the Ebola outbreak in 2014–16, covering the spread of the disease and the WHO/Mèdecins Sans Frontiers (MSF) response.

2. 'The Guinea Worm Eradication Programme'. **https://www.cartercenter. org/health/guinea_worm/**.

   Direct link to the Carter Foundation, and their discussion on the challenges they face in trying to eradicate this parasite.

3. 'A Brief History of Antibiotic Resistance: How a Medical Miracle Turned into the Biggest Public Health Danger of our Time'. **http://www.medicaldaily. com/antibiotic-resistance-history-373773**.

   A timeline of antibiotic resistance. Look at when the first concerns on this subject were raised, compared to when action started to be taken!

4. Sabib A.A. Jassim and Richard G. Limoges, 'Natural Solution to Antibiotic Resistance: Bacteriophages "The Living Drugs"'. *World J Microbiol Biotechnol* (2014) 30:8, 2153–70. **https://www.ncbi.nlm.nih.gov/pmc/ articles/PMC4072922/**.

   Article examining bacteriophages in greater detail, along with how they can be used to help combat antibiotic resistance in both humans and agriculture.

Discussion questions

7.1 Do you think that, with an ever-expanding human population, better transport links, and a shrinking pool of effective antibiotics, that an untreatable global pandemic is inevitable? What measures could be put in place to try to reduce the spread of an outbreak?

7.2 What barriers can you see to a large-scale trial of bacteriophages to treat bacterial infections? How would you go about designing a suitable trial to prove (or disprove) their efficacy?

GLOSSARY

Acquired resistance—The ability of bacteria to develop resistance to the common synthetic antibiotics used against them

Adjuvant—A substance which enhances or modifies the effect of other substances. May be used to make vaccines more effective

Aerobes—Organisms which require oxygen to survive

Agar jelly—A gel used to culture microorganisms

Agar plates—Petri dishes containing agar jelly, used to culture bacteria

Altered glycan theory—Changes in the glycans expressed on the surfaces of cells within the body trigger an inflammatory response—one theory explaining autoimmune disease

Amoebae—Single-celled organisms which move about using finger-like projections of protoplasma. They may be free-living or parasitic. Some are human pathogens

Amygdala—The area of the brain associated with emotional processing

Amyloid plaques—Clumps of abnormal proteins which accumulate in the central nervous system

Anaemic—Having too few red blood cells in the body. There are lots of different causes for anaemia, only some of which are caused by infectious disease

Anaerobes—Organisms which can survive without oxygen

Anthrax—A bacterial disease, caused by *Bacillus anthracis*, which can affect humans or animals. Anthrax can be fatal if untreated, and spreads easily from person to person

Antibodies—Large, Y-shaped proteins produced by the immune system in response to a specific antigen

Antigen—A substance foreign to the body, which induces an immune response, including the production of antibodies

Antigen presenting cells (APCs)—Dendritic cells, macrophages, or B-cells that present antigens to immature T-cells in the lymphoid tissue

Antimicrobial—Something that kills microorganisms, or otherwise stops their growth

Apoptosis—Programmed cell death

Aspergillosis—A fungal infection caused by fungi of the genus *Aspergillus*. Often affecting people with existing lung disease (such as asthma or chronic obstructive pulmonary disease), it may cause no symptoms at all, or may cause bleeding, fevers, chest pain, and breathlessness. If the fungus spreads elsewhere in the body it may cause widespread organ damage

Autoimmune disorders—A range of conditions in which the immune system targets and damages the organs of its host, instead of pathogens

Autopsy—The examination and dissection of a corpse, usually performed to determine the cause of death or to teach students about anatomy

Bacteriocidal—Describes an antibiotic (or other substance) which kills bacterial cells

Bacteriophage—A virus which has specialised in infecting prokaryotic cells

Bacteriostatic—Describes an antibiotic (or other substance) which inhibits bacterial reproduction

Basophils—Specialised leucocytes, releasing cytokines and eicosanoids

Biopsies—Samples of tissue taken from the body to be examined under the microscope

Biosecurity—The term used for a set of precautions that aim to prevent the introduction and spread of harmful organisms

Bipolar disorder—A psychiatric disorder characterised by periods of depression and periods of elevated moods known as mania

Body lice—Insect parasites which live on the human body. They may act as a vector for various infectious diseases

Botulinum toxin–Produced by *Clostridium botulinum*, this exotoxin can cause paralysis and death even in microscopic quantities

Broad spectrum–Used when referring to antibiotics which are effective against a wide range of different bacteria

Bubonic plague–A disease caused by the bacterium *Yersinia pestis*, causing fevers, headaches, vomiting, and the formation of large, painful swellings (called 'buboes') of the lymph nodes in the armpits and groin

Budding–The process of coating a virus with a viral envelope, consisting of viral spike proteins and a lipid bilayer

Buffers–Used to maintain the pH of a culture medium–in microbiology this is often to ensure that the microorganisms survive transport or storage

Cadaver–A dead body used for dissection

Capsid–The protein coat of a virus. Not all viruses have capsids

Carbolic acid–An aromatic compound, originally extracted from coal tar. A mild acid, it also has antiseptic qualities

Central nervous system (CNS)–The brain and spinal cord

Cerebral palsy–A group of permanent neurological disorders, caused by abnormal development or damage to different areas of the brain

Cestodes–Also known as tapeworms. There are several different tapeworm species which may infect humans, causing a variety of different symptoms

Childbed fever–Now known as puerperal fever, this refers to any infection caught by the mother at the time of giving birth

Chinese liver fluke–A parasitic flatworm, found in marshy areas

Chloramphenicol–An antibiotic with a relatively broad spectrum of action which works by inhibiting protein synthesis, blocking peptidyl transferase activity on the bacterial ribosome

Cholera–A disease causing severe watery diarrhoea, leading to death from dehydration and electrolyte imbalances if untreated. Caused by the bacterium *Vibrio cholerae*

Clostridium botulinum–The bacterium which produces botulinum toxin. It is anaerobic, meaning that it can exist in very low-oxygen environments

Common cold–An illness caused by rhinoviruses. Symptoms include increased mucus production from the nose, coughing, sneezing, and fevers

Complement cascade–A series of chemical reactions resulting in the production of various different proteins within the bloodstream. These proteins are directly involved in the immune response

Congenital toxoplasmosis–Occurs when a mother infected with *Toxoplasma* passes the infection on to her unborn child. This may lead to conditions such as epilepsy, hydrocephalus, jaundice, brain damage, learning difficulties, or cerebral palsy in the baby

Contact tracing–Tracking everyone an infected person may have had contact with, so they can be tested for the same disease, or monitored to see if they display symptoms

Creutzfeldt-Jakob Disease (CJD)–A prion disease caused by faulty PrP, resulting in progressive dementia and death

Cryptococcus–A genus of fungi which can occasionally cause fungal meningitis in people with suppressed immune systems

Crystal violet–A form of staining, taken up by the peptidoglycan in some bacterial cell walls

Culture medium–A substance in which microorganisms may be cultured

Culturing–Growing microorganisms in a laboratory setting

Cyst–A thin-walled, hollow cavity inside the body

Cysticercosis–A disease caused by formation of tapeworm cysts in the human body. Symptoms depend on where the cysts form

Cytokines–A number of substances, secreted by cells of the immune system, which effect a response on a cellular level

Cytotoxic–Describes a substance which causes damage or destruction to cells

Definitive host–The final host of a parasite's life cycle, often where it reproduces

Dendritic cells—Specialised phagocytes found in tissues which are exposed to the external environment—the nose, skin, stomach, intestines, and lungs. At certain stages in their lifecycle, they are shaped a little like dendritic neurons (nerve cells)

Dimorphic—Capable of taking on two separate forms

Dissection—The act of dismembering something (animal, plant, or human) to further study its anatomical structure

DNA—Deoxyribonucleic acid. A chain of nucleotides containing the genetic material essential for life

Droplet spread—The spread of a pathogen through droplets of body fluids (such as mucus from the airways)

Ectoparasites—Parasites which exist on the exterior of the body, rather than inside it

Eicosanoids—Signalling molecules made up of polyunsaturated fatty acids with a wide range of roles. They form crucial elements of signalling during the immune response

Electron microscope—A microscope which uses electron beams instead of light, generating extremely high magnification images up to 10,000,000 times

Elephantiasis—The most severe form of filariasis. Sufferer's limbs become severely deformed and chronically swollen, causing long-term pain and disability

Endocytosis—A method by which viruses enter their target cells. Glycoproteins on the viral envelope fuse with proteins on the host cell membrane, and the virus is then absorbed into the host cell, surrounded by a coat of the host cell's own membrane

Endoscopy—A way of examining the alimentary canal, using a flexible, fibre-optic camera inserted through the mouth or anus. Samples of tissue (biopsies) may be taken using tools attached to the endoscope

Endotoxins—Compounds produced by some bacteria, which form part of their cell membranes. When the bacterial cell undergoes lysis, these toxins can trigger a rapid immune response, making the infected person very unwell

Eosinophils—Specialised leucocytes, releasing cytokines and eicosanoids

Epidemic—A widespread outbreak of an infectious disease

Epidemiology—The study and analysis of patterns in health and disease

Epitope—A biological key on an antigen, which binds with the paratope on an antibody

Escherichia coli—Bacteria responsible for a number of different types of infection, depending on their subtype. Commonly causing gut or urine infections, *E. coli* are anaerobes, thriving in low-oxygen environments

Exocytosis—A method of exiting the cell used by both enveloped and non-enveloped viruses. The virus is encased in a vesicle, surfaced with the host cell's own lipid bilayer, and passed through the host cell's membrane into the extracellular space, where the vesicle ruptures, releasing the virus

Exotoxins—Compounds produced by some bacteria, which may be released at any time during the bacterium's life cycle. They are usually easily broken down by the immune system, but can be much more potent than endotoxins, and often work before the immune system has a chance to act against them

Facultative anaerobes—Bacteria which can survive in both aerobic and anaerobic environments

Faecal–oral route—Common route of transmission for infected disease. Poor hand hygiene allows transmission of pathogens from the anal tract to the mouth, where they are swallowed and may then cause infection

Fatal familial insomnia (FFI)—A prion disease caused by faulty PrP, resulting in progressive insomnia, insanity, and death

Filarial worms—A genus of nematodes which cause the disease filariasis

Filariasis—Infection of the lymphatic system by filarial worms. This may cause rashes, arthritis, and limb swelling, ultimately leading to the condition known as elephantiasis

Flukes—Another name for trematodes, taken from the Old English word for flounder

Free radicals—Atoms or groups of atoms with an odd number of electrons, which can cause damage to organic tissue by binding to cell membranes or DNA

Gelling agent—A substance added to culture medium to solidify it to a jelly-like state

Genetic code—The genes found on strands of DNA which encode the biological characteristics of the organism

Genital warts—Warty growths which form on the skin of the vagina, penis, or anus. They are caused by various subtypes of the human papilloma virus

Germ theory of disease—The idea that pathogens are responsible for causing infectious disease, rather than miasma or an imbalance of the humours

Gerstmann–Straussler–Scheinker disease—A prion disease caused by faulty PrP, causing coordination problems, dementia, and death

Giardia lamblia—Also known as Beaver fever, this protozoal parasite infects the intestines, causing diarrhoea, weight loss, and abdominal cramps

Haematopoietic stem cells—The stem cells which differentiate to form blood cells within the body

Headlice—Insect parasites which live on the human scalp. Athough they may cause itching, they are not a vector for other infectious diseases

Helminths—Parasitic worms. There are many different species of helminthic parasites

Herbalist—Someone who treats diseases using herbs and other plant-based cures. Many ancient societies relied heavily on herbalists to treat infectious disease

Herd immunity—The concept that even a vaccine that is not 100% effective at preventing disease in an individual can still be tremendously effective at reducing the rates of transmission through a group of people, if enough of the population are immunised

Hermaphroditic—Containing both male and female reproductive organs—many hermaphroditic organisms are capable of fertilising themselves

Histoplasma capsulatum—The fungus responsible for causing histoplasmosis in humans

Histoplasmosis—A fungal disease, which may cause respiratory symptoms and a flu-like illness. It may make immunosuppressed individuals very unwell

Hookworms—A genus of roundworms which sometimes use humans as their hosts

Human botfly—A fly which uses the human body as a nursery for its larvae. The larval form burrows under the skin, causing localised irritation

Human microbiome—The microorganisms which exist throughout our bodies, either causing us no harm or existing in symbiosis with us

Human papilloma virus (HPV)—A virus with multiple different subtypes, some of which are associated with cervical, penile, and anal cancers. Other subtypes are associated with genital warts. HPV is sexually transmitted

Humours—A part of early theories of disease in many different cultures. Ancient peoples believed that an imbalance of the humours (blood, phlegm, black bile, and green bile) was the cause of many different kinds of disease

Hydrocephalus—Excessive accumulation of fluid around the brain

Hyphae—Thread-like filaments which make up the structure of many fungi

Immunoassay—A way to quickly measure concentrations of different antigens or antibodies

Immunosuppression—Partial or complete suppression of the immune system. This may be deliberate (by the use of immunosuppressing drugs) or may occur due to one of many disease processes

Industrial revolution—The improvements in manufacturing and industry which took place in Europe during the eighteenth and nineteenth centuries

Inflammatory pathways—Chemical pathways which result in the release of chemicals causing inflammation within the body

Inoculating—The act of introducing infectious tissue into a non-infected individual

Intrinsic resistance—A pre-existing resistance to antibiotics, occurring when the natural characteristics of an organism make it resistant to one or more different antibiotics

Klebsiella—Bacterial genus with many different species and subspecies. Coloniser of human digestive tracts, where it may exist without causing disease—but also responsible for unpleasant infections in humans and animals

Leishmania trypanosomes—Parasitic protozoa responsible for causing leishmaniasis

Leishmaniasis—A disease caused by at least twenty different species of *Leishmania* trypanosomes, and spread by sandfly bites. It causes ulcerations of the skin and mucous membranes, and may also lead to liver damage and immunosuppression

Leprosy—An infection caused by the bacteria *Mycobacterium leprae* or *Mycobacterium lepromatosis*. This disease damages the nerves, which means that sufferers will often damage their bodies without realising it. Leprosy has been the subject of tremendous social stigma over the millennia, with people suffering from the disease often isolated in 'leper colonies' despite the fact that the disease is not very contagious

Leucocytes—White blood cells. There are multiple different subtypes of leucocytes

Lipid bilayer—The membrane of a cell, called a bilayer because it has two layers of phospholipids

Lyme disease—A disease caused by *Borrellia* bacteria and spread by bites from infected ticks. It causes a variety of localised and systemic effects, and can be very challenging to diagnose

Lymph nodes—Collections of lymphoid tissue found throughout the body, linked by lymphatic channels

Lymphatic—The lymphatic system is an important part of the human circulatory system, carrying a clear substance called lymph through lymphatic channels around the body

Lymphoid tissue—The organs and cells which make up the lymphatic system, including the spleen, lymph nodes, bone marrow, and thymus

Lymphokines—A subtype of cytokine, produced by lymphocytes. They act as signalling proteins between cells of the immune system

Lysis—The disintegration of a cell by disruption of its membrane

Lysosome—A vesicle containing hydrolytic enzymes designed to trigger cell lysis

Macrophage—A large phagocyte, a specialised leucocyte, which is dedicated to absorbing infected cells and foreign bodies, and which may remain stationary in the tissues of the body or circulate within the bloodstream

Major Histocompatability Complex I (MHC I)—Molecules found on the surface of cells within the body, which are used to help recognise 'self' and 'non-self' by the immune system

Major Histocompatability Complex II (MHC II)—Similar to MHC I, except that they are only found on the surface of antigen presenting cells, and the molecules are extracellular in origin rather than being produced by the cell itself

Malaria—An infectious disease caused by various subtypes of plasmodium parasites. It causes high fevers, anaemia, pain, diarrhoea, and is often fatal unless treated

Malathion—An insecticide used to treat infestations of several different ectoparasites

Mast cells—A type of specialised leucocyte, generally found in connective tissues and mucous membranes. They help to regulate the inflammatory response

Medical anthropologist—Someone who studies health and disease in relation to cultures and societies

Membrane attack complex—A structure formed on the surface of pathogens by the complement cascade, which aims to cause lysis of the pathogen

Meningitis—An infection of the lining of the brain and spinal cord (meninges). Meningitis may be bacteria, viral, or even fungal

Meningococcal septicaemia—Septicaemia resulting from an overwhelming infection by the bacteria that cause meningitis

Methicillin-resistant *Staphylococcus aureus* (MRSA)—A species of *Staphylococcus* which has become resistant to the antibiotic methicillin (and other antibiotics containing a beta-lactam ring). It is genetically different to other staphylococci, and can be very difficult to treat

Miasma theory—An ancient theory of infectious disease, stating that infectious diseases were caused by clouds of foul air called miasma

'**Missing self**'—The process of looking for certain markers (such as MHC I—Major Histocompatability Complex I) present on the surface of host cells

Mucous membranes—Tissues which line many bodily orifices, consisting of epithelial cells which secrete mucus

Multi-systems atrophy (MSA)—A prion disease caused by the prion α-synuclein. It is the first prion disease known which is not caused by the prion PrP

Mycelium—A network of connected hyphae

Mycobacterium tuberculosis—The curved, rod-shaped bacterium responsible for causing tuberculosis. It is encased in a thick, waxy coat, making it very resilient

Natural killer (NK) cells—Leucocytes which seek out and destroy damaged or compromised cells belonging to the host

Nematodes—Also known as roundworms—a group of worms which parasitise humans

Neomycin—An aminoglycoside antibiotic, binding to the 30S ribosomal subunit and inhibiting the ability of bacteria to make proteins. It is bactericidal, and is very effective against Gram-negative organisms—although only partially effective against Gram-positive ones

Neurodegenerative diseases—Progressive, degenerative diseases of the CNS. This includes prion diseases as well as diseases like Alzheimer's dementia and Parkinson's disease

Neutrophils—A form of phagocyte. They are short-lived and highly mobile

Norovirus—Also known as 'the winter vomiting bug', noroviruses cause severe diarrhoea and vomiting. They are very infectious, and can quickly infect large groups of people in hospitals, nursing homes, schools, or prisons. The infection is usually short-lived, but there is no cure or vaccine

Nucleotides—The basic structural units of nucleic acids like DNA; a compound made up of a nucleoside group bonded with a phosphate group

Obligate parasite—An organism which must parasitise others to survive. An obligate parasite cannot survive without its host

Oocysts—Cysts containing the eggs of an organism

Opsonisation—The targeting of a pathogen or infected cell for destruction by the innate immune system

Ova—Eggs

Pandemic—An epidemic of infectious disease that has spread across a large area

Paratope—A biological lock, configured in such a way that it will only bind with its corresponding epitope on an antigen

Pathogens—Organisms which can cause disease. They may be bacterial, viral, fungal, parasitic, or prions

Pathologist—A scientist (usually a doctor) who studies the causes and effects of disease, often by examining tissues and samples in a laboratory setting

Patient zero—The name given to the first detected case in a disease outbreak

Pattern recognition receptors—Specialised structures designed to aid the immune system in the recognition of pathogens and 'non-self' molecules

Penicillium—The genus of mould species used to create penicillin

Perforin—A cytolytic protein found within killer T-cells, which aims to trigger apoptosis in its target

Permethrin—An insecticide used to treat infestations of several different ectoparasites

Petri dish—A glass or plastic dish, used to culture microorganisms

Phagocytes—Leucocytes which are specialised in phagocytosis, engulfing and destroying pathogens or foreign matter

Phagocytosis—The absorption and digestion of bacteria or other material by phagocytes and protozoans

Phagolysosome—The name given when a phagosome and lysosome combine

Phagosome—A vesicle used to engulf a pathogen. This is then fused with a lysosome to destroy its contents

Pin-worms—A genus of roundworms which may use humans as their hosts

Pit latrine—A primitive toilet, in which the human waste falls into a deep pit

Plagues—The term used by ancient people to describe epidemics

Plasma cells—Mature B-cells which produce a single antibody

Plasmodium—The genus of organisms responsible for causing malaria

Pneumocystitis pneumonia (PCP)—A lung infection caused by the fungus *Pneumocystis carinii*, causing fever, breathlessness, and other symptoms of pneumonia

Polymerase chain reaction (PCR)—A way to quickly make multiple copies of a segment of genetic code (DNA or RNA), using an enzyme called polymerase

Prions—Pathogenic, misfolded proteins. They may arise spontaneously or be transmitted from exposure to infected tissue. Prions are not living organisms, but cause a variety of progressive, fatal, neurological disorders in humans

Proglottids—The body segments of a cestode. Each proglottid is capable of producing and fertilising its own eggs

Prokaryotes—Single-celled organisms which do not possess a membrane-bound nucleus, mitochondria, or any other membrane-bound organelles

Protozoa—Single-celled microorganisms belonging to the kingdom Protista

Pseudopodia—Foot-like structures, used by protozoa to help them move and catch prey

Pubic lice—Insect parasites which live in human pubic hair. They may cause itching, but do not act as a vector for other infectious diseases

Public health—The science of disease prevention and tracking, prolonging life and promoting health improvements

Quarantine—The act of isolating an infected person or population from others, preventing the spread of an infectious disease

Rat bite fever—An infection caused by two different bacteria (*Streptobacillus moniliformis* or *Spirillum minus*), causing fever, arthritis, rashes, and swelling of the lymph nodes

Respiratory distress syndrome—A collection of symptoms caused by the lungs being unable to supply the body with adequate oxygen. Symptoms include breathlessness, wheeze, drowsiness, and ultimately loss of consciousness

Rhinoviruses—Common, tiny viruses which cause colds and upper respiratory tract infections

Rigors—Symptoms of a high fever—shaking, shivering, feeling cold despite a high body temperature, and sweating profusely

RNA—Ribonucleic acid. A biological molecule essential in coding, decoding, regulation, and expression of genes

Roundworm—A parasitic worm, which may be a member of a number of different species

Salvarsan—A drug developed in the early twentieth century, used to treat syphilis

Scabies mite—The mite responsible for causing the infectious disease scabies, which causes itching and irritation to the skin

Schistosomes—A group of trematodes responsible for a disease known as schistosomiasis

Schistosomiasis—An infection caused by schistosome parasites, causing bloody diarrhoea or bleeding from the bladder. Longer-term infection causes damage to the organs, as the eggs released by the parasite become engrained within the body's own tissues

Schizophrenia—A psychiatric disorder with a variety of presenting symptoms. Sufferers may experience hallucinations, delusions, disordered thoughts and speech, and abnormal social behaviour

Scolex—The head unit of a cestode. The arrangement of hooks and suckers on the scolex can help to identify the species of worm

Sepsis—A state in which the organs of the body begin to sustain damage due to overwhelming infection and the body's own immune response

Septa—Porous walls through which nutrients may pass in a fungus. Singular: septum

Septicaemia—A situation in which large numbers of bacteria have entered the bloodstream and are circulating throughout the body

Serotypes—Distinct variations between otherwise similar microorganisms, often determined by looking at the antigens presented on the surface of the pathogen

Sexually transmitted infections (STIs)—A variety of infectious diseases spread by sexual contact. They may be bacterial, viral, or even parasitic

Species barrier—The differences between species which make it difficult for a pathogen to spread from one species to another

Spleen—An organ in the abdomen which forms part of the lymphatic system, aiding in the storage and maturation of B-cells, and the filtering of red blood cells

Spontaneous generation—The ancient belief that all life could be created spontaneously from non-living matter

Spores—Parcels of genetic material, used by fungi to reproduce

Staining—A technique used to help identify bacteria, involving using different washes and dyes on microscope slides

Strain—A genetic variant or subtype of a pathogen

Streptomycin—An antibiotic in the aminoglycoside group. It works by inhibiting the ability of 30S ribosomal subunits to make proteins, which kills the bacteria

Sulfonamides—A class of antibiotic used extensively during the early twentieth century to treat a wide range of infections. Issues with allergy, quality control, and toxicity led to the creation of regulatory authorities around the world, which still exist today

Symbiotic relationship—A relationship that is mutually beneficial to both parties

Syphilis—A sexually transmitted infection causing, amongst other symptoms, skin rashes, joint pains, madness, and eventually death. It is treatable with antibiotics

Tapeworm—A parasitic worm, which may be a member of a number of different species

Tetracycline—The first antibiotic in the tetracycline family, this antibiotic works by binding to the 30S ribosomal subunit and inhibiting bacterial protein synthesis, thereby killing the bacterium

Trematode—Parasitic flatworms which may infect human hosts

Trichomonas vaginalis—A protozoa which infects the vagina, causing itching, foul-smelling vaginal discharge, pain on urination, and increased risk of HIV infection

Trypanosomes—A group of single-celled parasitic protozoa with flagella, which move with a corkscrew-like motion

Typhoid—A disease caused by *Salmonella typhi*, causing high fevers, bloody diarrhoea, and often death if untreated

Typhus fever—An infectious disease caused by *Rickettsia prowazekii* and spread by body lice

Undefined medium—A culture medium which has unknown quantities of amino acids. It contains all of the elements that many bacteria and fungi need to survive

Vaccination—The process of exposing the immune system to a weakened or dead form of a pathogen, thereby activating T and B memory cells and rendering the person vaccinated immune to the diseases caused by that particular pathogen

Vector—An agent which carries or spreads a pathogen, such as a mosquito or a flea

Vesicle—An enclosed bubble of fluid surrounded by a lipid bilayer

Viral shedding—The stage of viral reproduction in which new viruses are released from the cell in which they have been produced

Virus envelope glycoproteins—Proteins which may stud the surface of a viral envelope. These may help the virus to enter its target cells

Whipworm—A parasitic worm, which may be a member of a number of different species

WHO—The World Health Organization. A worldwide organisation which works in over 150 different countries to improve health and healthcare

Whole genome sequencing (WGS)—The process of determining the entire DNA sequence of an organism's genome at a single time

INDEX

A

acquired resistance 140–1, 146
adaptive immune response 77, 78–83, 108–9
aerobic bacteria 28, 72
agar media 55–6
agglutination 81
allergies 48–9, 77, 88, 115
amyloid plaques 52
anaerobic bacteria 28–9, 56, 72
ancient world
    medical knowledge 6–9, 110, 119
    medical systems, India 7
    Silk Road, spread of parasites 4–6
    traditional Chinese medicine (TCM) 7–8
anthrax 6, 18, 28, 113
antibiotics 29, 73, 87–93, 147
    bacterial resistance 140–5
antibodies 80–3, 108–9
antifungal medications 95–8
antigen presenting cells (APCs) 79
antigens 75–6, 77–8, 79, 80–1
antimicrobial properties
    botfly maggot secretions 51
    drugs, mechanisms of action 87, 88–9, 91–2, 99
    human body secretions 71
    traditional herbal treatments 7, 8
antiparasitic medications 98–101
antiseptic surgery 18, 87
antiviral drugs 38, 93–4
apoptosis 33, 79
asthma, and parasitic infections 48–9
autoimmune disorders 83, 95
Ayurvedic medicine, India 7

B

bacteria
    culturing 55–6, 144–5
    discovery 12–13
    effects of antibiotics 87, 88–9, 91–2
    growth patterns 56–7
    humans as hosts 24–7, 72–3
    reproduction and genetics 23–4
    resistance development 140–1, 146
    structure and classification 27–9, 58–60
    toxin release 19
bacteriophages 32, 146–7
Baltimore classification (viruses) 34–5

basophils 77
B-cells 79–80, 82
biosecurity 42, 62–6
Black Death (bubonic plague) 6, 9–12
body lice 50–1
botfly, human 51
broad-spectrum antibiotics 90, 92, 93, 99
budding (viruses) 33

C

cancers 19–20, 25–6, 27, 45
    chemotherapy effects 95
cannibalism 54
capsids, viral 31, 32, 35
Carter Foundation 136–7
causes of disease
    ancient beliefs 6–9
    germ theory 16–18, 121
    humours 7, 8, 9
    medieval theories 9–10
    recent discoveries 18–20, 26–7
    see also diagnosis; pathogens
cestodes (tapeworms) 5, 45–7
chemical defences 71, 75
childbed fever 16–17
children
    deaths from disease 2, 101, 118–19
    diarrhoea incidence 27, 35
    gut flora development in infancy 72–3
    parasite infestations 45, 100
    vaccination 105, 115–19
Chinese liver fluke 5–6
Chinese medicine, traditional 7–8
cholera 28, 120–2
cities, living conditions 4, 8, 120–1, 122
CJD (Creutzfeldt–Jakob Disease) 52, 54
commensal organisms 72–3
common cold 35, 137, 143
complement cascade 75–6, 81
contact tracing 65, 123, 126, 128
cowpox vaccination trials (Jenner) 111–13
culture of pathogens
    differential media 56
    growing media 55–6
    Koch's postulates 18
    minimal media 55
    nutrient media 55

selective media 55–6
    for strain diagnosis 144–5
cysticercosis 46
cysts (of parasites) 45, 46, 137–9
cytokines 43, 73, 75, 79–80

D

deaths
    decline, and causes 91, 109, 119
    historical extent 1–2, 9, 86, 108
    of laboratory workers 64–5
    mortality rate in childbirth 16–17
    in pandemics, WHO statistics 38, 130, 132, 134
    parasite infections 45, 101, 103
    viral infections, annual numbers 35, 36
β-defensins 71
dementia 52
dendritic cells 77–8
diagnosis
    new diseases 53–5, 65, 130–1
    parasite infestation 46–7, 60–1, 98–9, 103
differential media 56
digestive tract
    microbiome 24, 72–4
    parasites 43, 45–6
domestic animals 3–4, 137–9
droplet spread 35, 37, 127
drug development
    antifungal drugs 95, 97–8
    antiparasitic drugs 98–9
    antivirals 93
    penicillin and other antibiotics 89–91, 141, 145, 147

E

Ebola 65, 132–4
ectoparasites 50–1
eicosanoids 75
electron microscopy 31, 61, 65
elephantiasis 47
endocytosis 32, 34
endotoxins 19
envelope, viral 31, 32, 33, 37–8
eosinophils 77
epidemics 4
    cholera outbreaks 120–2
    historical, in Europe 9–12
    new diseases 129–30
    smallpox, and use of variolation 111